Jörg Haus

Optical Sensors

Related Titles

Brown, T. G., Creath, K., Kogelnik, H., Kriss, M. A., Schmit, J., Weber, M. J. (Eds.)

The Optics Encyclopedia

Basic Foundations and Practical Applications. 5 Volumes

2004
ISBN: 978-3-527-40320-2

Gross, H. (Ed.)

Handbook of Optical Systems

Volume 1: Fundamentals of Technical Optics

2005
ISBN: 978-3-527-40377-6

Udd, E. (Ed.)

Fiber Optic Sensors

An Introduction for Engineers and Scientists

2006
ISBN: 978-0-470-06810-6

Göpel, W., Gardner, J. W., Hesse, J. (Eds.)

Sensors Applications

5 Volumes

2005
ISBN: 978-3-527-29565-4

Baltes, H., Fedder, G. K., Korvink, J. G. (Eds.)

Sensors Update 13

2004
ISBN: 978-3-527-30745-6

Jörg Haus

Optical Sensors

Basics and Applications

WILEY-VCH Verlag GmbH & Co. KGaA

The Author

Dr. Jörg Haus
Ringstr. 18
35614 Aßlar
Joerg_Haus@T-Online.de

■ All books published by Wiley-VCH are carefully produced. Nevertheless, authors, editors, and publisher do not warrant the information contained in these books, including this book, to be free of errors. Readers are advised to keep in mind that statements, data, illustrations, procedural details or other items may inadvertently be inaccurate.

Library of Congress Card No.: applied for

British Library Cataloguing-in-Publication Data
A catalogue record for this book is available from the British Library.

Bibliographic information published by the Deutsche Nationalbibliothek
The Deutsche Nationalbibliothek lists this publication in the Deutsche Nationalbibliografie; detailed bibliographic data are available on the Internet at http://dnb.d-nb.de

© 2010 WILEY-VCH Verlag GmbH & Co. KGaA, Weinheim

All rights reserved (including those of translation into other languages). No part of this book may be reproduced in any form – by photoprinting, microfilm, or any other means – nor transmitted or translated into a machine language without written permission from the publishers. Registered names, trademarks, etc. used in this book, even when not specifically marked as such, are not to be considered unprotected by law.

Cover   Adam Design, Weinheim
Typesetting   Toppan Best-set Premedia Limited
Printing and Binding   Betz-druck GmbH, Darmstadt

Printed in the Federal Republic of Germany
Printed on acid-free paper

ISBN: 978-3-527-40860-3

To Birgit and Felix

Contents

Preface *XI*

Introduction *1*
References *3*

Part One The Optical Sensor Construction Kit *5*

2 **Light Sources** *7*
2.1 Important Properties of Light Sources *7*
2.2 Thermal Light Sources *8*
2.3 Line Sources *12*
2.4 Light Emitting Diodes (LEDs) *14*
2.5 Lasers *18*

3 **Photodetectors** *27*
3.1 Photomultipliers *27*
3.2 Photodiodes *30*
3.3 Other Detector Types *31*
3.4 Imaging Detectors *33*
3.5 Detector Noise *35*

4 **Optical Elements** *37*
4.1 Optical Materials *37*
4.2 Mirrors, Prisms and Lenses *39*
4.3 Dispersive Elements: Prisms and Gratings *42*
4.4 Optical Filters *43*
4.5 Polarizers *44*
4.6 Optical Fibers *46*
4.7 Modulators *49*
References *53*

Optical Sensors: Basics and Applications. Jörg Haus
© 2010 WILEY-VCH Verlag GmbH & Co. KGaA, Weinheim
ISBN: 978-3-527-40860-3

Part Two   Optical Sensors and Their Applications   55

5   Eyes: The Examples of Nature   57
5.1   The Compound Eyes of Insects   57
5.2   Nature's Example: The Human Eye   58

6   Optical Sensor Concepts   63
6.1   Switches   63
6.1.1   Light Barriers   63
6.1.2   Rain Sensor   65
6.2   Spatial Dimensions   66
6.2.1   Distance   66
6.2.2   Displacement   71
6.2.3   Velocity   75
6.2.4   Angular Velocity   83
6.3   Strain   86
6.4   Temperature   90
6.5   Species Determination and Concentration   93
6.5.1   Spectrometry   94
6.5.2   Polarimetry   98
6.5.3   Ellipsometry   102
6.5.4   Refractometry   104
6.5.5   Particle Density and Particle Number   106
6.5.6   Fluorescence Detection   111
6.6   Surface Topography   114
6.6.1   Chromatic Confocal Sensors   115
6.6.2   Conoscopic Holography   117
6.6.3   Multiwavelength Interferometry (MWLI)   118
6.6.4   White-Light Interferometry   119
6.6.5   Near-Field Optical Microscopy   122
6.6.6   Contouring: Structured-Light Techniques   123
6.6.7   Concepts: Cross-Correlation Analysis and 2D Fourier-Transform Techniques   125
6.7   Deformation and Vibration Analysis   127
6.7.1   Laser Vibrometers   127
6.7.2   Speckle-Pattern Interferometry   129
6.7.3   Holographic Interferometry   132
6.8   Wavefront Sensing and Adaptive Optics   135
6.9   Determination of the Sun Angle   138
6.10   Determination of Age   140
           References   143

Part Three Optics and Sensors at Work: A Laser-Spectroscopic Experiment *147*

7 **The Measurement Problem** *149*

8 **The Physical Principles behind the Experiment** *151*

9 **Spectroscopic Setups of the Experiment** *155*
9.1 The Single-Step Approach *155*
9.2 The Two-Step Approach *158*
     References *165*

Summary *167*
Glossary *169*
Index *173*

Preface

During the collection of data and bibliographic information for an article that covered the history of a particular optical sensor family, I became amazed at the degree of sophistication that was possible even before sensor designers knew about highly sensitive, highly dynamic, and highly integrated optoelectronic components. Going back into history, it was interesting to see how researchers had very pragmatically found a way to solve technical problems with a component set that – at least in our modern understanding – was still somewhat simple.

Today, having reached the "century of the photon," gathering knowledge about all modern-day components that can be incorporated into an optical sensor is of utmost importance for everyone who is occupied with optical sensor design. Also, there are numerous optical sensors for almost every measurement task, particularly when a physical contact between sensor and measurement object is undesired or even impossible. Despite this variety of components and sensors, no introductory textbook – at least to the author's knowledge – has been published yet that comprehensively covers the basics and possible fields of application of optical sensors.

Thus, the motivation of this book is to give an overview over the toolbox of light sources, optical elements, and detectors employed in optical sensor concepts. By describing a variety of sensor principles between the ubiquitous light barrier and sophisticated high-resolution surface-scanning interferometers, it shall provide a starting point into the world of optical sensors both for students of, for example, electrical engineering, optical technologies or automation technologies, and for professionals in all industries in which optical sensors may be of importance. It is intended to collect information that could otherwise only be obtained by the extensive study of component or sensor data sheets, or by studying multiple sources and (often very specialized) monographies on single sensor concepts.

As always, this project would not have been possible without the help of everyone who supported me in the realization of this manuscript: First of all, I would like to express my sincere thanks to Dr. Andreas Thoß of Wiley-VCH, who appreciated the idea of a manuscript on optical sensors and who finally brought me in contact with Mrs. Anja Tschörtner of the physics department of his publishing house. To her, and also to Dr. Christoph von Friedeburg, Associate Publisher

Optical Sensors: Basics and Applications. Jörg Haus
© 2010 WILEY-VCH Verlag GmbH & Co. KGaA, Weinheim
ISBN: 978-3-527-40860-3

Physics Books, I am deeply grateful for providing me with valuable information, help, and support throughout the entire writing and editing process. Last, but not least, I owe my deepest thanks to my wife Birgit and my little son Felix for their love, encouragement, and understanding during the long time it took me to finish this manuscript.

November 2009 *Jörg Haus, Aßlar*

Introduction

The German Agenda "Optical Technologies for the 21st Century" describes optical technologies as enabling technologies for other technical fields and their applications in the future [1], which has just recently been confirmed by a study of the German Ministry of Education and Research (Bundesministerium für Bildung und Forschung, BMBF, [2]). Already at the very beginning of this twenty-first century, optical sensors play a vital role in virtually every technical application: from simple light barriers to complicated white-light interferometers, from automotive rain sensors to high-resolution scanning near-field optical microscopes, there are sensors to measure a wide variety of measurement properties.

Optical sensors offer various advantages over their electronic and/or mechanical counterparts. Apart from their wide dynamic range and lower noise levels, optical sensors react on any change in the measurement properties literally with the speed of light. Also, the noncontact nature of the optical measurement avoids all systematic errors that come with tactile techniques. Consider the measurement of brake path lengths in vehicle testing: a well-established technique employs a combination of a so-called fifth wheel with an incremental encoder. To calculate the measurement result with sufficient measurement uncertainty, the wheel diameter must be known with the same precision. Here, the wheel slip will introduce systematic measurement errors. This is why the majority of automobile manufacturers relies on sensors that determine dynamic measurement properties from the optical cross-correlation between an image of the road surface and an optical grating [3].

Much effort is taken to miniaturize optical sensors and to integrate them into electrooptical chip designs. Not only the wide availability of megapixel-size imaging sensors, for example, for mobile phones, notebook computers, and miniaturized surveillance cameras, but also the ubiquitous optical computer mouse with its highly integrated sensor chip and optics are well-known examples. It is interesting to note that this high integration of state-of-the-art optical systems can also be found in biological systems, for example, in the optical system that is most important for all of us: the human eye. In a first approximation, it works like a simple hole camera with a layer of photoreceptors in its image plane. However, the "inverted" design of the retina seems quite strange from an optical designer's point of view: the light receptors are arranged behind several tissue layers and

Optical Sensors: Basics and Applications. Jörg Haus
© 2010 WILEY-VCH Verlag GmbH & Co. KGaA, Weinheim
ISBN: 978-3-527-40860-3

should therefore actually be subject to significant light scattering. It has just recently been revealed that the so-called Müller cells are the crucial functional elements inside the retina: They form an effective image-guiding optical fiber plate and, thus, transport the received image without significant losses across the cell layers directly to the light receptors [4].

Regardless of their working principles and purposes, all optical sensors have a common set of components: light sources, photodetectors, and optical components to guide the light in between. Somewhere along this light path, there will be the measurement object whose static and dynamic properties will determine the signal on the photodetector and, thus, the sensor reading. In order to understand the working principles of different optical sensors, it is therefore necessary to learn about the possible types of illumination and detection hardware, and about the various optical components that may be used. This will be discussed in Part One of this book. In Part Two, setups of different optical sensors are presented. They will be arranged according to the physical or chemical quantities that the sensors are intended to measure. Due to the large variety of optical sensors available today, this collection will not claim completeness, but will illustrate the state-of-the-art in optical sensor technology. Part Three will shed some light on the role of optical sensors in physics – it describes a laser-spectroscopic experiment from the author's physical background and its elaborate experimental setup. Some of the components employed in this setup have also been discussed in Part Two. The third part tries to give some brief introduction to the physical principles behind the experiment, but it does in no way want to ask too much from the reader.

The idea of this book is to bridge the gap between classical optical textbooks and monographies on particular optical sensor technologies. While the former cover all aspects of physical optics with sometimes only basic descriptions of applications, the latter delve deep into singular applications without detailed descriptions of the physical and optical basics. This leaves room for an introductory textbook, which is both aimed at students of all relevant engineering and scientific disciplines, and at interested professionals in all industries in which optical sensors are employed. It is intended to cover both aspects to an extent that gives the reader not only an idea about how different optical sensors work, but also about the possible fields of application and their respective limits.

No textbook that covers a topic like this can avoid to give references to commercially available sensors and sensor systems. The reader should note that any company and brand names are properties of their respective owners. Although the author has and had affiliations with at least one of the companies mentioned in the text, all examples are mere snapshots of the market situation and do not express recommendations for particular products or manufacturers. Similarly, the specifications listed up in the various tables are intended to give "typical" sensor data. The reader is strongly encouraged to do some research on his or her own, for example, on the Internet, in case an optical sensor shall be employed for a measurement task.

References

1 Siegel, A. and Litfin, G. (eds) (2002) *Deutsche Agenda Optische Technologien für das 21. Jahrhundert*, VDI-Technologiezentrum, Düsseldorf.
2 Bundesministerium für Bildung und Forschung (BMBF) (ed.) (2007) *Optische Technologien – Wirtschaftliche Bedeutung in Deutschland*, BMBF, Bonn, Berlin.
3 Delingat, E. (1976) *LEITZ Mitt. Wiss. Techn. Bd. VI*, **7**, 249–257.
4 Franze, K., Grosche, J., Skatchkov, S.N., Schinkinger, S., Foja, C., Schild, D., Uckermann, O., Travis, K., Reichenbach, A., and Guck, J. (2007) *Proc. Natl. Acad. Sci. U S A*, **104** (20), 8287–8292.

Part One
The Optical Sensor Construction Kit

No matter what an optical sensor is good for and how it is designed in detail, it always consists of

- a light source,
- a photodetector, and
- a number of optical elements.

The object that the sensor points to is a part of the measurement process in the sense that it "processes" the incident light in such a way that the quantity to be measured can be extracted from the photodetector signal (Figure 1.1). However, the particular property to be measured, environmental conditions, target price, and many other parameters define the application-specific setup and the adequate set of components.

This part of the book gives a summary of the different kinds of light sources, of the wide variety of commercially available detector elements, and provides examples of optical elements that can be employed to guide and manipulate light.

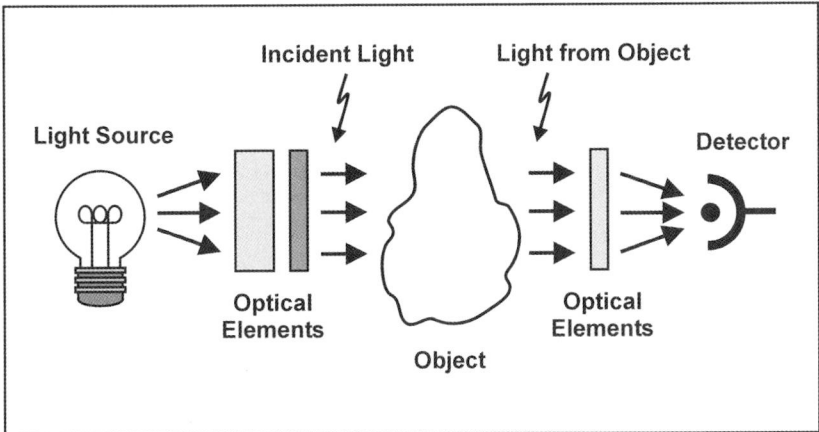

Figure 1.1 An optical sensor.

Optical Sensors: Basics and Applications. Jörg Haus
© 2010 WILEY-VCH Verlag GmbH & Co. KGaA, Weinheim
ISBN: 978-3-527-40860-3

2
Light Sources

The light source is the very heart of an optical sensor. It not only provides the "medium" through which information is transferred, but it may also become a component of the detection circuit itself, for example, by a modulation of the emitted light.

A light source is characterized in terms of, for example, its emission spectrum, degree of coherence, radiant intensity, power consumption, lifetime, and all other parameters decisive for the respective application. Therefore, this chapter starts with a summary of these parameters before it reports on the most important light sources to date. Due to the physical nature of their light generation process, they can be divided up into thermal sources and line sources. Lasers, although line sources as well, will be treated in a separate section due to their very particular type of radiation.

2.1
Important Properties of Light Sources

One of the most important properties of a light source is its emission spectrum. This spectrum may be either wide and offer a great variety of wavelengths, or it may be narrow, monochromatic and, thus, only suited for very particular applications. In the visible part of the electromagnetic spectrum, the shape of the emission spectrum determines the light color perceived by the human eye, which is generally represented by a coordinate in the two-dimensional CIE (Commission Internationale de l'Éclairage) color space. The German standard DIN 5033 describes hue, saturation, primary and combination colors in the frame of this color space. Table 1.1 gives a compilation of perceived colors and their respective wavelength ranges. Without going any further into this comparably complicated matter, an optical sensor must follow the general design rule to contain a light source/detector pair for which the emission spectrum of the former matches the spectral sensitivity of the latter as closely as possible.

The property that determines the maximum signal that the detector will yield is the emitted light power. This property is characterized by several parameters.

Optical Sensors: Basics and Applications. Jörg Haus
© 2010 WILEY-VCH Verlag GmbH & Co. KGaA, Weinheim
ISBN: 978-3-527-40860-3

Table 1.1 Perceived colors and corresponding wavelength ranges (approximate values).

Color	Wavelength range (nm)
Violet	380–450
Blue	450–500
Cyan	500–520
Green	520–565
Yellow	565–590
Orange	590–625
Red	625–740

Table 1.2 Photometric units.

Property	Symbol	Unit	Conversions
Luminous intensity	I_V	1 Candela (cd)	1 cd = 1 lm/sr
Luminous flux	Φ	1 Lumen (lm)	1 lm = 1 cd sr[a]
Luminous energy	Q_V	1 lm s	
Luminance	L_V	1 cd/m²	
Illuminance	E_V	1 Lux (lx)	1 lx = 1 lm/m²
Luminous emittance	M_V	1 Lux (lx)	

a) sr, steradian, unit for solid angle.

Photometric units in the visible part of the spectrum are based on the SI base unit of the luminous intensity, the candela (cd):

- **Luminous flux:** Light power perceived by the human eye.
- **Luminous energy:** Light energy perceived by the human eye.
- **Luminance:** Brightness of a light source. The smaller a light source, the brighter it is (at identical luminous intensities).
- **Illuminance:** Intensity of light incident on a surface.
- **Luminous emittance:** Intensity of light emitted from a surface.

Table 1.2 presents a summary of these properties, their common symbols, their SI units, and how they can be converted among each other.

2.2
Thermal Light Sources

Thermal or incandescent light sources emit light as a result of their temperature. Every object with an absolute temperature T emits a continuous spectrum which is described by the spectral power density $I_{e\lambda}$ according to Planck's law of blackbody

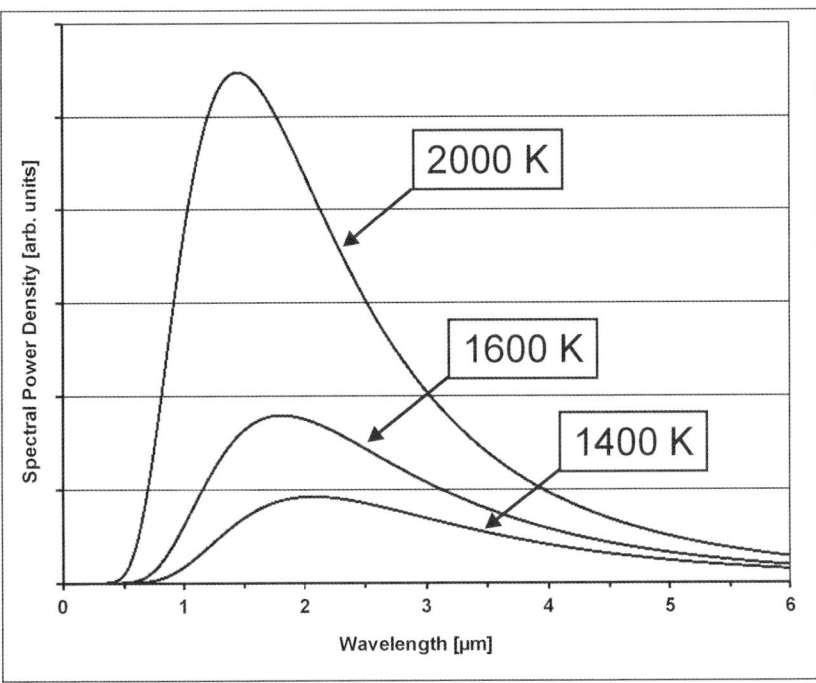

Figure 1.2 Planck's law. Spectrum of a thermal light source for different temperatures.

radiation. With $h = 6.626 \times 10^{-34}$ J s (Planck constant), $c = 2.9979258 \times 10^8$ m/s (light speed in vacuum), λ: wavelength, $k = 1.38 \times 10^{-23}$ J/K (Boltzmann constant), the emission spectrum, that is, $I_{e\lambda}$ as a function of λ is described by [1]

$$I_{e\lambda} = \frac{2\pi hc^2}{\lambda^5}\left(e^{\frac{hc}{\lambda kT}} - 1\right)^{-1} \tag{1.1}$$

Since a blackbody is only an idealization for a real glowing body, the right-hand side of Eq. (1.1) has to be multiplied with the spectral emissivity, $\varepsilon(\lambda)$, according to Kirchhoff's law. The spectral emissivity denotes the efficiency of emission or absorption relative to the blackbody, for which ε equals 1 for all wavelengths. Figure 1.2 shows the calculated spectra for absolute temperatures of 1400 K, 1600 K, and 2000 K.

The continuous spectrum reaches a maximum and approaches zero for higher wavelengths. With increasing temperatures, the maximum shifts toward smaller wavelengths. It can be found from Eq. (1.1) that this shift of the wavelength of maximum spectral power density, λ_{max}, is inversely proportional to the absolute temperature of the glowing body, and it is described by Wien's law [1]:

$$\lambda_{max} \cdot T = 2.898 \times 10^{-3} \text{ m K} \tag{1.2}$$

Also, by integrating Eq. (1.1), one finds that the total emitted radiant power, I_e, from the unit surface is determined only by the body's absolute temperature [1]:

$$I_e = \sigma \cdot T^4 \tag{1.3}$$

with $\sigma = 5.67 \times 10^{-8}\,\text{W}\,\text{m}^{-2}\,\text{K}^{-4}$. Equation (1.3), known as the Law of Stephan and Boltzmann, means that by doubling the temperature of a glowing body, its total radiant power will increase by a factor of 16.

The following three examples may illustrate these equations:

- The spectrum of sunlight has a shape that, although containing absorption structures from the Sun's photosphere and from the Earth's atmosphere, can very well be described by Eq. (1.1). As the emission maximum is at around 500 nm, the Sun's surface temperature can be estimated to be about 5800 K.

- Molten steel with an absolute temperature of 2000 K has its emission maximum at 1.4 µm wavelength, which is in the infrared region of the electromagnetic spectrum. Equation (1.2) may, therefore, very well be employed in noncontact measurements of extremely high temperatures. By the way: the reason why the color of melted steel is indeed perceived as yellow-white is that the combination of its emission spectrum and the spectral sensitivity of the human eye yields a maximum in this part of the visible spectrum.

- The core of the human body has a temperature of about 37 °C = 310 K. This corresponds to a maximum in the emitted thermal radiation at a wavelength of about 9.3 µm. This is the basis for the already wide-spread radiation thermometers for the measurement of body temperatures.

One of the easiest embodiments of a radiating blackbody is an electrically heated, glowing piece of wire or a metal filament. In the presence of oxygen, such a filament would oxidize almost immediately. The filament is therefore surrounded by a glass vessel filled with an inert gas, for example, with a mixture of nitrogen and argon, or with even heavier noble gases like krypton or xenon. This is the basic principle of all incandescent lamps.

The maximum temperature of the filament is only limited by the melting point of the material, usually tungsten. At the usual temperature of about 2800 K, however, the optical efficiency of a light bulb – the ratio between the radiant intensity in the visible spectrum and the applied electrical power – is rather poor and typically around 5%. Their luminous efficacy, the ratio between luminous flux and electrical power, is around 10 lm/W. The lifetime of a light bulb is mainly limited by the rate by which material is evaporated from the filament. When it reaches its lifetime of around 1000 h, the filament is weakened so far that it breaks, and then the light bulb must be replaced.

Higher filament temperatures and, thus, higher effectivities are achieved with tungsten halogenide lamps. Their filaments are made of tungsten as well, but the bulb is filled with iodine or bromine vapor. When material is evaporated from the filament, it reacts with the halogen atoms and forms tungsten–hexahalogenides (WI_6 or WBr_6, respectively). These molecules dissociate again when they come in

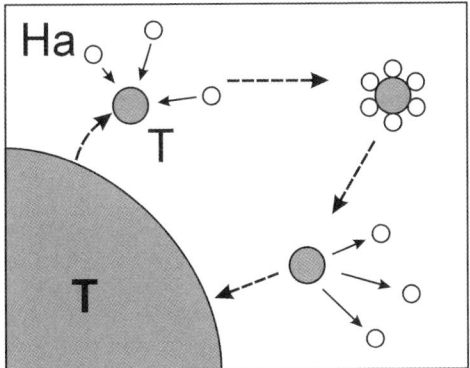

Figure 1.3 Tungsten halogen lamp: cyclic regeneration of the filament material (schematic). T: tungsten atom, Ha: halogen atom.

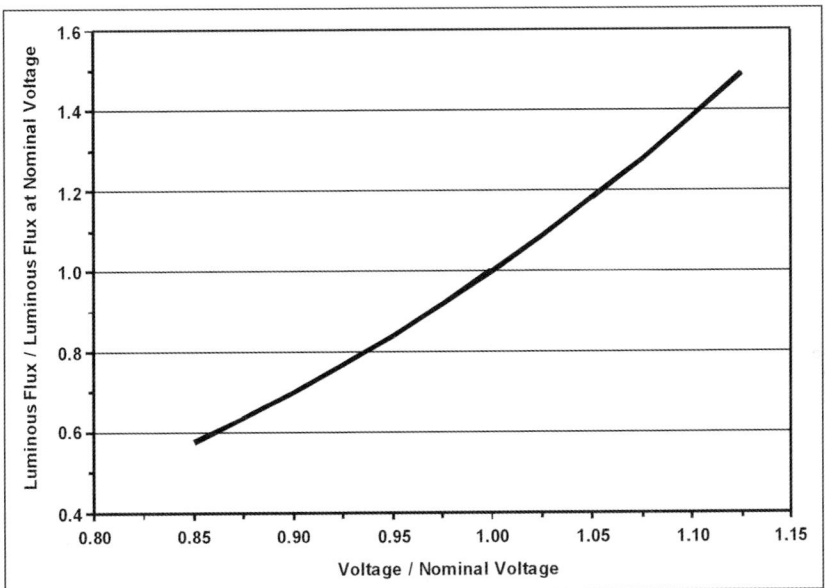

Figure 1.4 Halogen lamp, luminous flux as a function of supply voltage.

contact with the hot filament surface. There, the tungsten atoms are adsorped back again (Figure 1.3). This cyclic regeneration of the filament material allows higher filament temperatures of up to 3500 K. With a lifetime of around 2000 h, the luminous efficacy of halogen lamps is around 25 lm/W.

Incandescent lamps have strongly nonlinear characteristics. With increasing supply voltage U, the luminous flux increases with $U^{3.4}$, but the lifetime of the lamp decreases with U^{-16} [2]. Figures 1.4 and 1.5 show these relations, where the relevant properties are expressed relative to their respective nominal values.

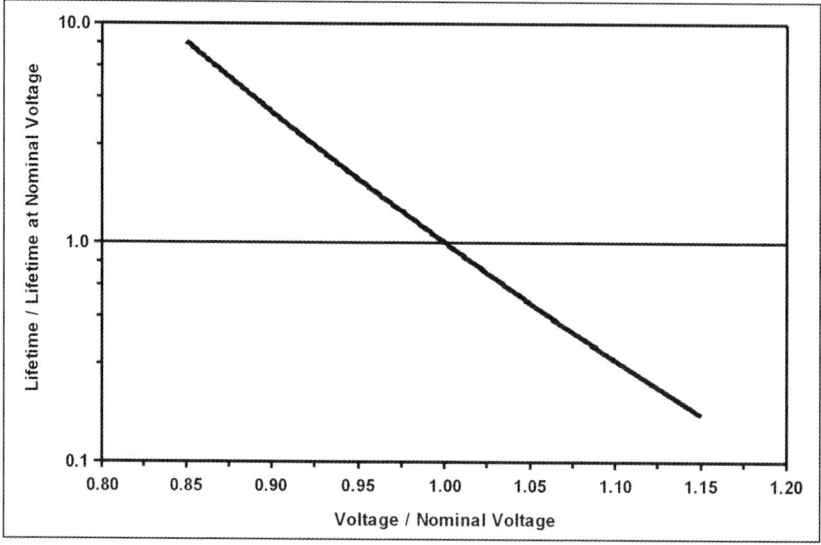

Figure 1.5 Halogen lamp, lifetime as a function of supply voltage.

The graphs reveal that a voltage reduction of 5% from its nominal value leads to a reduction of the luminous flux of about 15%, but also to a 1.5 times longer lifetime. This trade-off is often used to increase lamp lifetimes when they are required to operate for a long time, for example, when they cannot be replaced with ease.

2.3
Line Sources

A discharge in a low-pressure gas does not emit a continuous spectrum, but a spectrum in which the intensity is concentrated in discrete and narrow wavelength intervals. When viewed with a spectrograph, these intervals become visible as single lines in the spectrum, which is therefore denoted by a line spectrum.

The discrete distribution of emitted wavelengths is the consequence of the discrete energy levels of atomic orbits. When two atoms collide, or when atoms collide with free electrons in the discharge, atomic electrons will be excited from one discrete atomic energy level, characterized by its principal quantum number n, to another level with higher principal quantum number m. From this level, E_m, the electron drops to a lower level, E_n ($n < m$), which is not necessarily the level from which the excitation process had started. This relaxation process usually takes about 10^{-8} s, and the energy difference is radiated off as a quantum of light with frequency v (Figure 1.6).

The energy of this quantum, $h \cdot v$, can be calculated according to [3]

$$h \cdot v = E_m - E_n = 13.6 \text{ eV} \cdot Z^2 \cdot \left(\frac{1}{n^2} - \frac{1}{m^2} \right) \tag{1.4}$$

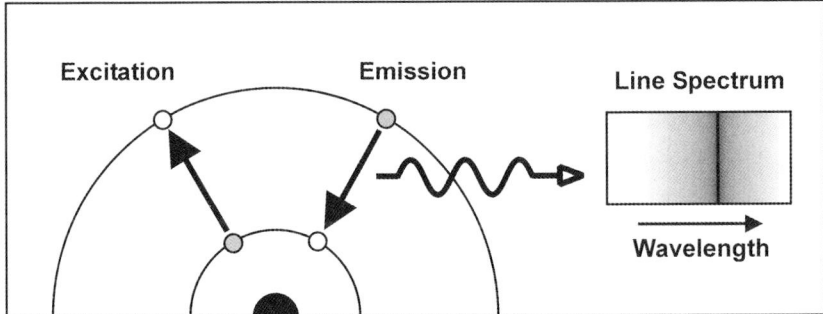

Figure 1.6 The physical principle behind a line source: electron jumps between discrete atomic energy levels.

Here, ν is the frequency of the emitted radiation and Z the nuclear charge. The wavelength of the emitted radiation can easily be calculated using the wave equation $\lambda = c/\nu$. Every transition described by Eq. (1.4) is represented by one discrete line in the emission spectrum. It is interesting to note that without the factor Z^2, this equation does not only result from Bohr's semiclassical atomic model, but also from the quantum-mechanical treatment of the hydrogen atom. In the quantum-mechanical formalism, however, more correction factors than a simple Z^2 must be added. For an introduction into this formalism, see, for example, Ref. [4].

As their emission is concentrated in a limited number of discrete wavelength intervals, discharges in a low-pressure gas do generally not emit white light, but show color effects:

- Neon has several emission lines in the visible spectrum, but also a dominant line at 632.8 nm. This is why neon tubes emit pink light.
- Sodium (Na) lamps emit yellow light because of the Na D lines, a narrow-spaced line doublet at 589.0 nm (D_2) and 589.6 nm (D_1). During power-up, however, they emit red light due to the neon added to the gas in the bulb.
- Mercury has a strong emission line at 253.7 nm. Mercury tubes, also called blacklight tubes, emit ultraviolet (UV) light. The ubiquitous fluorescent lamps are manufactured from these blacklight tubes by coating the insides of their glass vessels with phosphor substances that convert the UV to visible light. Their spectrum is white, but there are also cool-white types with color temperatures, that is, the temperature of a blackbody with the same color impression, of more than 4000 K.

The width of the emission lines depends on temperature and pressure. Higher temperatures lead to broader velocity distributions of the atoms and, thus, to broader frequency distributions, the so-called Doppler broadening. Higher pressures, on the other hand, increase the collision rates between the atoms and, thus, lead to a decrease in the lifetimes of the upper states, which also leads to a line broadening. Xenon or mercury lamps, for example, are clearly line sources. High

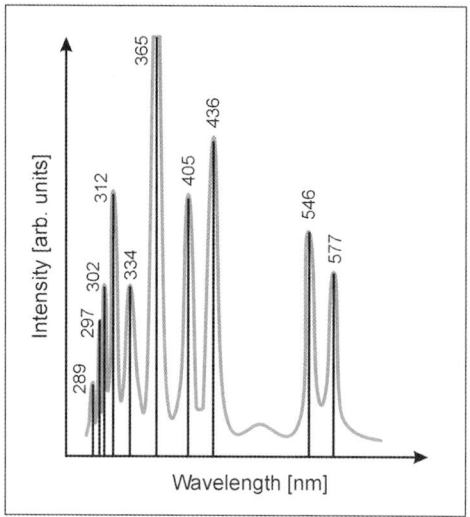

Figure 1.7 Spectrum of a mercury short-arc lamp. Black solid lines: line center positions.

temperatures and pressures in the glass vessels, however, broaden the lines so extremely that their line spectra turn into continuous white-light spectra. Figure 1.7 shows the spectrum of a mercury short-arc (HBO) lamp: while the positions of the emission lines are well-defined by quantum mechanics, the gaps between the lines will be filled completely due to the pressure-induced increase in linewidth. The result is a continuous, white-light spectrum. As there are several ultraviolet and blue lines in the spectrum, however, these lamps have a visual emission that is slightly shifted to the blue. The ultraviolet lines qualify them as powerful ultraviolet light sources, for example for fluorescence microscopy. A typical 50 W HBO lamp reaches a luminous flux of 11 000 lm, a luminance of 70 kcd/cm^2, and a mean lifetime of 200 h [5].

Metal-halide lamps are close relatives to mercury-vapor lamps. They contain a high-pressure mixture of mercury, argon, and metal halides. While argon helps to ignite the discharge, the metal halides determine the emitted spectrum. With color temperatures of around 4000 K, these lamps reach luminous efficacies of more than 100 lm/W, that is, 10 times more than an incandescent lamp. Their lifetime, on the other hand, can be as long as 20 000 hours (see, e.g., [6]), which makes them considerably more cost-effective than incandescent lamps.

2.4
Light Emitting Diodes (LEDs)

When free atoms condensate to regular structures and form solid matter, their sharp energy levels become disturbed so far that they form quasicontinuous energy bands. These bands are still separated by forbidden zones, so-called gaps, which can be crossed by electronic transitions. The uppermost occupied band is denoted by the valence band, and the lowermost free band the conduction band. An electronic transition from the lower edge of the conduction band to the upper edge of the valence band will lead to the emission of a light quantum, and its energy will equal the width of the energy gap, E_G:

$$h \cdot v = E_G \tag{1.5}$$

The actual light source in a semiconductor material is the contact region between an n-doped and a p-doped host crystal, the pn-junction. Throughout the junction, the number of free charge carriers, negatively charged electrons from the n-side and positively charged holes from the p-side, is low due to recombination processes. These processes lead to an electrical field across the junction that will finally stop the recombination current. In this equilibrium state, any external electrical field will drive additional electrons and holes into the junction where they will recombine. These recombinations result in emission of light. This is, however, just a simplified explanation of the involved processes. A comprehensive theoretical description of the energy band model of solid-state physics and the description of electronic transitions in its frame will be far more complicated. Interested readers are, therefore, referred to Ref. [7] for further details.

Equation (1.5) implies that light emitted from a pn-junction has a narrow-band spectrum. Indeed, LEDs have monochromatic spectra with wavelengths from the ultraviolet to the infrared regions, with emission bandwidths of around 50 nm, or even less. In the visible range, typical high-power LEDs dissipate between 3 W and 5 W of electrical power and transform it into light with an efficiency of about 20%. As this light power is produced in a small volume, particular care has to be taken for the thermal management: Unless the dissipated power is not conducted away from the chip efficiently enough, the LED's lifetime will be reduced significantly. If however, all parameters remain within their nominal ranges, manufacturers state average lifetimes of several 10 000 hours until the luminous flux has decreased down to 70% of its initial value. Table 1.3 shows a collection of typical performance data. Parameters for the ultraviolet LED are taken from [8], those for the visible types from [9], and those for the infrared emitter from [10].

Table 1.3 Typical parameters of monochromatic LEDs[a].

Color	Wavelength (nm)	Luminous flux or radiometric power	Luminous efficacy or efficiency	Bandwidth (nm)
Ultraviolet	385	350 mW	0.12	10
Blue	470	35 lm	12 lm/W	25
Green	530	100 lm	30 lm/W	35
Amber	590	40 lm	45 lm/W	14
Red	627	60 lm	60 lm/W	20
Infrared	850	100 mW	0.03	40

a) Luminous flux/radiometric power: maximum possible values.

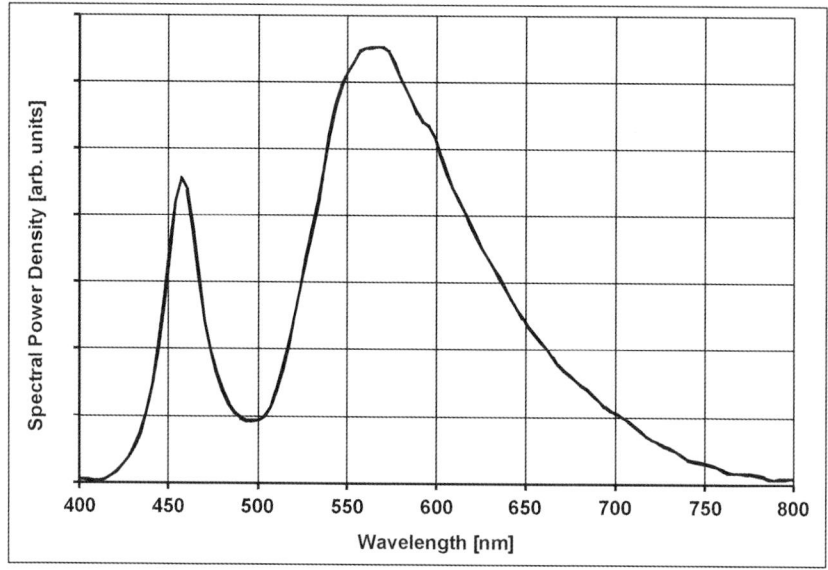

Figure 1.8 Spectrum of a white-light LED. Excitation peak at 470 nm, broadband emission with maximum at around 570 nm.

Because of the extremely small exit apertures with sizes of 0.01 to 10 mm^2, light emission from an LED shows considerable, diffraction-introduced divergence. For an optical designer, forming the light output from an LED is an important task, and manufacturers offer types with different radiation patterns, for example, Lambertian emitters that follow a cosine law for the radiated luminous intensity (θ represents the angle measured relative to the forward radiation direction):

$$I(\theta) = I(\theta = 0) \cdot \cos(\theta) \tag{1.6}$$

White-light LEDs are usually based on short-wavelength LEDs that are covered with a layer of photoluminiscent material. Their spectra show the narrow excitation peak in the blue or ultraviolet wavelength range, but also a broad emission covering the entire visible range (Figure 1.8). The spectral structure of this broad emission determines the color temperature of the emitted light. Today, optical designers can choose from cool-white, neutral-white, and warm-white LEDs (Table 1.4). To name just a few applications, white-light LEDs are used in torch lights, daytime running lights in automobiles, in microscopes and in general lighting. One major drawback is the low intensity emitted in the green part of the visible spectrum. Thus, tungsten halogenide lamps still have advantages over white-light LEDs whenever it is important to capture the true color of an object.

Organic Light Emitting Diodes (OLEDs) are based on recombination processes in thin organic layers with thicknesses between 80 nm and 200 nm. Today, these devices are used in displays, but to date, they do not yet have any meaning for lighting applications. There are, however, predictions that OLEDs may reach lumi-

Table 1.4 Typical parameters of white-light LEDs[a].

Type	Color temperature (K)	Luminous flux (lm)	Luminous efficacy (lm/W)
Cool-white	6500	120	40
Neutral-white	4100	120	40
Warm-white	3000	100	33

a) Luminous flux: maximum possible values.

Table 1.5 Typical parameters of superluminescent diodes (SLEDs)[a].

Wavelength (nm)	Spectral width, FWHM[b] (nm)	Output power (mW)
680	8.5	5.0
840	50.0	25.0
980	30.0	30.0
1020	100.0	5.0
1400	65.0	2.0
1550	40.0	2.0

a) Output power: measured at the exit of a single-mode fiber pigtail.
b) FWHM, full width at half maximum.

nous efficacies of more than 100 lm/W. The current status of OLED developments is reviewed in Ref. [11].

Superluminescent diodes (SLEDs) are LEDs manufactured from semiconductor materials with high optical gains as they are actually used in diode lasers (see below). In contrast to lasers, however, there is no resonator that feeds radiation back into the material so that there are basically two counterpropagating waves inside the active zone. The light from an SLED will thus be quite intensive, but it will have an LED-like bandwidth and therefore low coherence. In contrast to lasers, there will be no speckles across a spot that is illuminated by an SLED. With increasing temperature, the output power of an SLED will decrease significantly. As an example, this may result in a reduction of the output power by a factor of 10 when the temperature increases from 25 °C to 50 °C [12]. SLEDs are preferred light sources in many fiber-optical applications and are therefore also available in "pigtailed" packages, with a short piece of optical fiber through which the light exits. Table 1.5 shows a collection of typical SLED specifications [13].

With the advent of high-power LEDs, particularly those in the UV and blue spectral ranges, eye safety has become an important issue. Until 2008, both LEDs and lasers were classified according to EN 60825-1, but this standard overestimated the potential hazards of LEDs. By the time that this manuscript was compiled, LEDs are not contained in the most recent version of this standard any more. LEDs are now classified according to IEC 62471:2006 (Photobiological safety of lamps

2.5
Lasers

LASER is the acronym for Light amplification by stimulated emission of radiation. Stimulated emission is a quantum-mechanical process in which long-lived excited atomic states are quenched by radiation. These states are required to have lifetimes several orders of magnitude longer than the usual 10^{-8} s. Thus, it comes to a population inversion in the sense that more atoms are in excited states than in their ground states. As a consequence, one incoming quantum of light generates one additional quantum of outgoing light, and this results in light amplification.

Similar to an audio amplifier, the device becomes a signal source by feeding the output signal back to the input of the system. This feedback is usually accomplished by an optical resonator, a set of two parallel mirrors that contain the amplifying medium. The two mirrors have reflectances of more than 99% so that the light passes the medium more than once. The light waves traveling forward and backward inside the optical resonator interfere and form a standing wave. As a whole number q of half waves must fit into the resonator length, L, the distance between the two mirrors, the frequency v_q of the radiation is:

$$v_q = q \cdot \frac{c}{2L} \tag{1.7}$$

Equation (1.7) describes the so-called longitudinal modes of the laser. The mode spacing, the distance between two modes in frequency space, amounts to $c/2L$ and is constant for a given resonator length. In the visible and ultraviolet ranges of the spectrum, the existence of longitudinal modes is a necessity for a domination of stimulated over spontaneous emission. As the emission is localized only on narrow frequency intervals, the laser emission is extremely monochromatic: If the entire visible spectrum was spread over a distance of 10 km, the width of one laser mode would correspond to the width of a human hair.

Figure 1.9 illustrates how resonator modes and emission line profile determine the spectral output of a laser. The medium facilitates emission over the frequency range of the Doppler-broadened transition line, but only at the positions of the resonator modes. Also, due to losses induced by diffraction, scattering, and absorption, a minimum gain, the so-called laser threshold, is required for a mode to oscillate.

The resonator mirrors must perfectly be aligned to ensure that the light waves be reflected exactly along the optical axis. Thus, this optical axis is the direction in space into which light is emitted by the laser. Due to diffraction, however, a laser beam still has a divergence in the range of a few millirads.

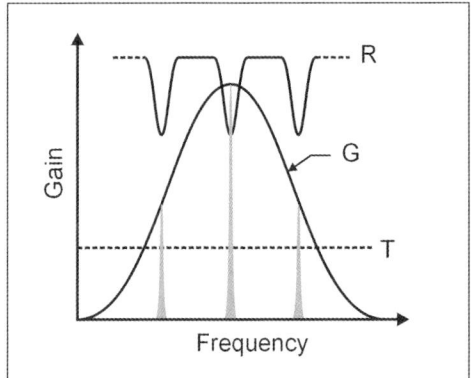

Figure 1.9 Output of a laser. R: resonator modes, G: Doppler-broadened profile of the emission line, T: laser threshold.

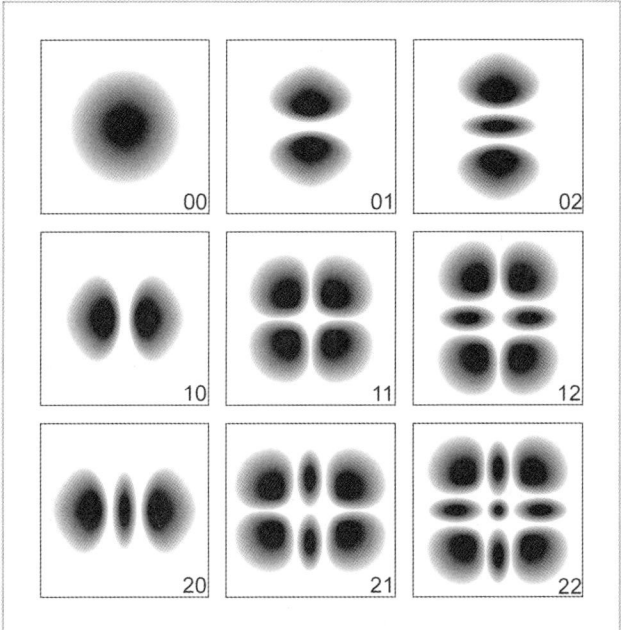

Figure 1.10 Transversal (TEM_{mn}) modes of a laser with rectangular mirrors. The numbers in the single images indicate the indices m and n, respectively.

The intensity distribution across the beam profile is determined by the transversal modes. These modes are usually described by the symbol TEM_{mn} (i.e., transversal electromagnetic), where m and n are small integers and describe the number of horizontal and vertical lines with minimal intensity in the beam profile. Figure 1.10 shows the first nine transversal modes of a laser with rectangular

Table 1.6 Laser safety classes according to EN 60825-1.

Class	Description
1	Eye-safe under all operating conditions. Visible or invisible radiation.
1M	Safe when viewed with the naked eye, but potentially hazardous when viewed with optical instruments. Visible or invisible radiation.
2	Eye-safe when viewed accidentally under all operating conditions, but not safe when viewed deliberately for longer than 0.25 s. Visible radiation.
2M	Eye-safe when viewed accidentally as long as the natural aversion response is not overcome, but potentially hazardous when viewed with optical instruments. Visible radiation.
3R	Low-risk, but potentially hazardous radiation. Class limit: five times the Class 1 limit for invisible, or five times the Class 2 limit for visible radiation.
3B	Potentially dangerous radiation. The maximum output of a continuous-wave laser into the eye must not exceed 500 mW. Hazard to eye or skin. Viewing of the diffuse reflection is safe.
4	Very dangerous radiation. Viewing of the diffuse reflection is dangerous. Radiation is capable of setting fire to materials.

mirrors. The preferred transversal mode is the fundamental TEM$_{00}$ mode: its Gaussian-shaped beam profile has minimum diffraction losses, and the wavefronts are almost spherical [14].

As laser light is strongly concentrated in one beam that hardly changes its diameter over long distances, a radiant intensity of, for example, 1 mW and a beam diameter of 3 mm will amount to a light intensity (the radiant power per unit surface) of $I \approx 300 \,\text{W/m}^2$. This is in extreme contrast to the light of a 100 W light bulb: as it is distributed almost isotropically, its light intensity decreases with distance following an inverse-square law. With the bulb's optical efficiency of 5%, a detector placed at a distance of 1 m will measure a radiant intensity of only $0.4 \,\text{W/m}^2$. This clearly shows the necessity to formulate and respect laser safety regulations that oblige laser manufacturers to classify each of their lasers according to the laser safety standard EN 60825-1 for continuous-wave radiation (Table 1.6).

Another important property of a laser is its coherence. The temporal coherence of a light source is determined by its spectral bandwidth and describes how well the emitted wavetrains interfere. In order to overlap, the upper limit for the spatial distance between two wave trains is their length, the so-called coherence length, l_{coh}. It is related to the linewidth of the light source, $\Delta \nu$, and the coherence time, Δt_{coh}, according to

$$l_{\text{coh}} = c \cdot \Delta t_{\text{coh}} = \frac{c}{\Delta \nu} \tag{1.8}$$

As an example, one emission line from a discharge lamp has a linewidth of around 5 GHz, which results in a coherence length of about 6 cm. On the other hand, a helium-neon (HeNe) laser with a linewidth of only 5 MHz has a coherence length of 60 m.

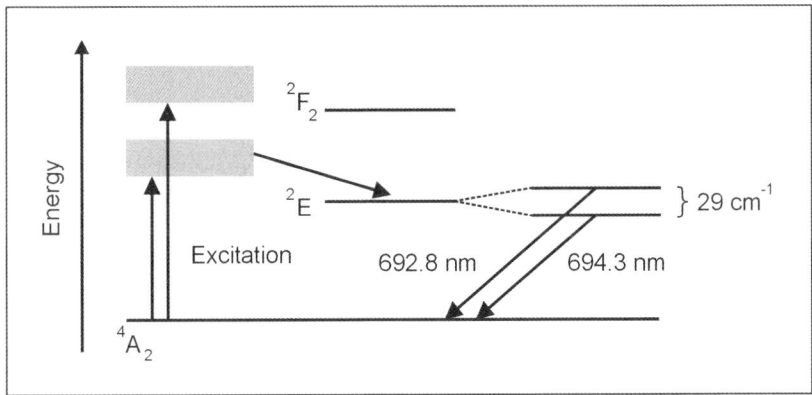

Figure 1.11 Excitation scheme of a ruby laser. The symbols denote particular properties of the electron states in the ruby crystal.

Spatial coherence describes the surface area, S, that a light source with circular area A_S can illuminate coherently with the light cone of its 0th diffraction order. In a distance r from the light source, S equals to

$$S = \frac{\lambda^2 \cdot r^2}{A_S} \tag{1.9}$$

Spatial coherence can thus be improved by increasing the distance to the source, and by reducing its size. This is the reason why point sources are always more coherent than those with large areas, and one practical way to increase spatial coherence is limiting A_S with light stops.

The amplifying medium of a laser may be solid, liquid, or gaseous. Solid-state lasers are based on ion-doped crystals or glasses, for example, Cr^{3+}, Nd^{3+}, Ho^{3+}. Laser transitions start and end at the inner energy levels of these ions, and because these levels are shielded from influences by the electrical fields of the host crystals, the transitions are already small by nature.

Although solid-state lasers reach high radiant powers, their efficiency is typically something around 0.1% which renders them unsuitable for sensor applications. However, one of the best-known examples of this laser family shall be described here: the first laser ever, built by Theodore M. Maiman in 1960. The active medium is a ruby rod, whose Al_2O_3 matrix contains typically 0.05% of Cr^{3+} ions. Ruby has two absorption bands of about 100 nm width which are located at 404 nm (blue) and 554 nm (green) wavelength. These bands are the cause of the crystal's red color. A flash lamp is a convenient way to excite both absorption bands (Figure 1.11). The first relaxation step ends in a split Cr^{3+} level, and both of these levels are metastable with an extremely long lifetime of about 3 ms. The higher level has a slightly smaller population so that the gain on the transition at 692.8 nm is smaller than that on the second transition at 694.3 nm. Thus, the laser starts to oscillate on the line with the larger wavelength. The smaller wavelength, however, may be selected with dispersive elements. As three energy levels participate in the

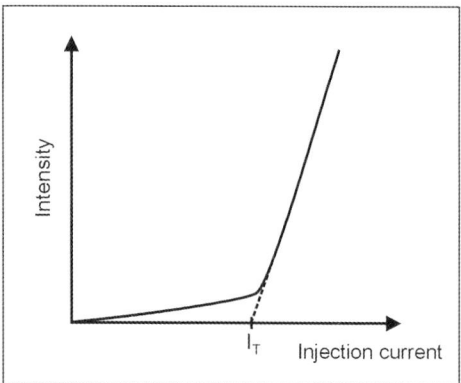

Figure 1.12 Dependence of laser diode intensity on the injection current. I_T: threshold current.

laser process, this is called a three-level system. In these systems, continuous-wave (cw) operation is usually difficult to achieve because of the high laser threshold. However, extremely high pulse energies can be reached.

Because of their size and their effectivity, semiconductor or diode lasers are more convenient to use in sensor systems. In principle, they are highly-doped LEDs with donor and acceptor concentrations of more than $10^{19}\,\mathrm{cm}^{-3}$. As in LEDs, electron–hole recombination processes lead to the emission of light, but due to the high doping, population inversion builds up in the pn-junction. As the excitation mechanism is based on the injection of charge carriers across the junction, these lasers are also called injection lasers.

The laser resonator is the crystal itself so that its faces have to be perfectly parallel and polished. Semiconductor materials have large refraction indices: for silicon, it is about 3.5 times larger than that of air, but still smaller than for germanium. The material will therefore have a reflectance of more than 30% against air. Special laser diodes, for example, single-mode lasers that oscillate on only one longitudinal laser mode, not only have a simple pn-junction, but also a significantly more complex setup. For a detailed discussion about this, the interested reader is referred to Ref. [14].

Diode lasers have two important characteristics. The first is the dependence of the light intensity on the injection current (Figure 1.12). For low-injection currents, the emitted intensity increases slowly. However, a threshold current, I_T, exists above which the intensity increases strongly. Due to this strong increase, highly stable currents are required for applications in which highly stable intensities are required. For currents smaller than I_T, the diode laser has the spectral properties of an LED, above the threshold current, the spectral width decreases due to increasing light amplification (Figure 1.13).

While the first commercial diode lasers had emission wavelengths in the infrared and near-infrared regions, mass-produced diode lasers with red, green, and blue emission wavelengths have become available in the meantime. The green diode laser is a complicated setup in which a neodymium-doped yttrium-

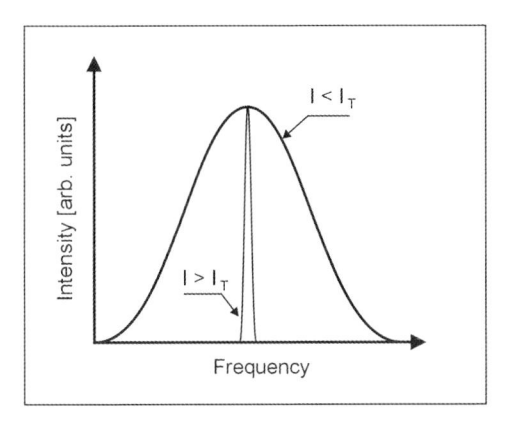

Figure 1.13 Diode laser spectrum for currents below and above the threshold current, I_T.

Table 1.7 Typical parameters of diode lasers in the visible and infrared regions.

Wavelength (nm)	Typical radiant power (mW)	Threshold current (mA)	Beam divergence (°)
405	45	35	8 × 19
635	10	60	10 × 30
670	60	80	10 × 30
780	100	30	10 × 30
850	100	50	10 × 30
905	300	50	25 × 40

aluminum garnet (Nd:YAG) crystal is pumped with a diode laser at around 800 nm and emits at 1064 nm. A so-called nonlinear crystal doubles the light frequency and therefore yields an output at one-half of this wavelength at 532 nm. The optical system can be miniaturized so far that it fits into the case of a standard laser diode. Blue laser diodes are based on special semiconductor structures of gallium nitride that emit wavelengths between 400 and 420 nm. These blue lasers can be focused onto a spot of one-fourth of the size of that of a red laser. For the storage of data, this means a drastic increase in the amount of data that can be written onto storage media like CD-ROMs or DVDs.

Table 1.7 shows a collection of typical specifications of diode laser modules of different wavelengths. Data for the blue-violet diode are adapted from [15], and those for the red and near-infrared types from the product portfolio of one large distributor [16]. Comparable data for green laser diodes cannot be given because they are offered as complete modules with additional optics and electronic drivers. The laser diode package generally contains a photodiode that has one common electrode with the laser diode. This photodiode receives the laser radiation emitted backward from the laser crystal and is used as a sensor for the intensity stabilization of the laser.

Diode lasers reach efficiencies of as high as 50% because electrical energy is directly transformed into radiation. For single laser diodes, the radiant power can reach some 10 mW, but there also exist high-power modules powerful enough to weld sheet metal. Due to this direct energy conversion, diode lasers can easily be modulated with frequencies up to the gigahertz range. There are also single-mode versions that emit only one longitudinal laser mode. These lasers are interesting light sources for spectroscopy.

Diode lasers are very small, and so are their exit apertures. These apertures generally have a rectangular profile, so diffraction effects cause high beam divergence and an elliptical beam profile. To form a parallel laser beam, a collimating lens is required, and circular beam profiles demand further optical corrections. Particularly for single-mode lasers, however, the collimating lens must be coated in order to avoid reflections back into the laser. This feedback light would also be amplified in the laser's pn-junction and would cause the laser to chaotically jump between different laser modes. Due to the random nature of these mode jumps, there would be no way to compensate this with stabilizing feedback loops. As additional protections against backreflections, there are also so-called optical isolators that prevent light from travelling into the direction of the laser.

Like with LEDs, a thermal management setup has to ensure that the dissipated power is efficiently conducted away from the laser. Also, diode lasers are extremely sensitive to electrical discharges that may damage or even destroy them. In case of damage, the laser diode may still work as an LED with the same emission wavelength as the laser, but with a larger spectral linewidth (Figure 1.13).

A very special version of the diode laser is the vertical-cavity surface-emitting laser (VCSEL). Laser light is emitted perpendicular to the plane of the pn-junction, which makes it possible to run tests already in the wafer level of the production process. As a VCSEL has a higher exit aperture, its beam divergence is smaller, and its light output can more efficiently be coupled into an optical fiber. The resonator mirrors are formed by layers of different refractive indices, so-called Bragg reflectors. Their reflectivities can reach 0.99, and thus the laser threshold is significantly lower than that for the pn-type diode lasers. However, their radiant power is smaller.

The best-known type of a gas laser is the HeNe laser. Its active medium is an approximately 7:1 mixture of ^{3}He and ^{20}Ne in an evacuated glass vessel with electric contacts, the discharge tube. Additionally, about 9% of ^{22}Ne are added with an emission profile that is shifted by about 1 GHz from that of the other isotope. This results in a total gain profile that is about twice as wide as it would be with only one isotope. Intensity variations during mode sweeps across this profile, for example, by scanning or due to temperature changes, will thus be significantly smaller.

The pumping process is initiated by starting a glow discharge in the tube. In the first excitation step, free discharge electrons excite the helium atoms. As there is a coincidence between the excited helium states and the relevant metastable states of neon, second-order collisions then transfer excitation energy to the neon

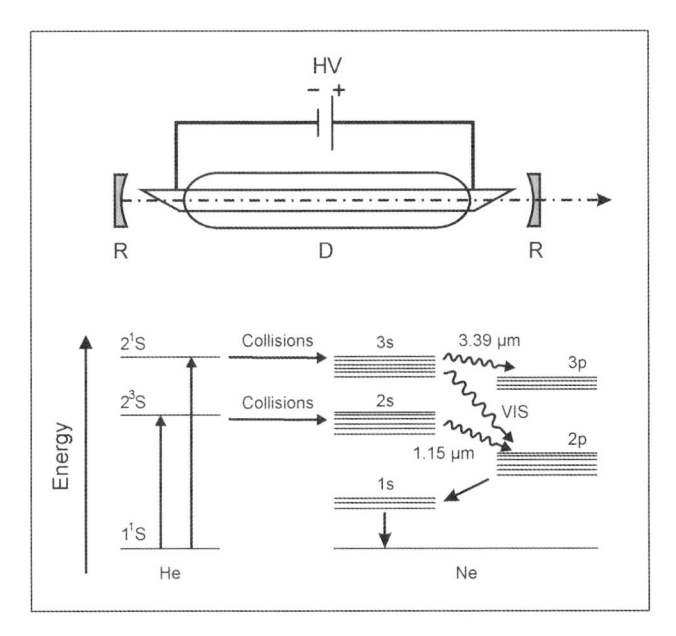

Figure 1.14 Schematic setup of a HeNe laser (a) and excitation scheme (b). R: resonator mirrors, D: discharge tube, HV: high-voltage supply. VIS: transitions with visible laser lines.

atoms. This can be written in the form of an equation, where the asterisk denotes an excited atom:

$$\text{He}^* + \text{Ne} \rightarrow \text{He} + \text{Ne}^* \tag{1.10}$$

The population inversion builds up by an increase in the population of the 2s and 3s levels of neon, each with a lifetime of about 100 ns (Figure 1.14). From these states, quantum-mechanical selection rules allow transitions only to the 2p and 3p levels, respectively, with lifetimes that are by about a factor of 10 smaller. These s–p transitions are the actual laser transitions: the 2p state decays within 10 ns to the 1s state and, by collisions with the walls of the discharge tube, further to the Ne ground state. As this last step is mandatory for a cw operation of the laser, the discharge tube contains a narrow glass capillary along the optical axis.

The strongest laser line is the infrared line at 3.39 µm, and a second infrared line is at 1.15 µm wavelength. The visible laser lines (Table 1.8, data adapted from [17]) can be selected with suitable resonator mirrors or with additional dispersive elements, for example, gratings or prisms. Radiant powers are in the milliwatt range, but there are also versions with up to 50 mW output.

Although the size of the discharge tube prevents a significant miniaturization of the laser, HeNe lasers are easy to handle and, therefore, well suitable for sensor systems with no extreme downsizing demands. Although these lasers require high-voltage power supplies, it takes only currents of a few milliamperes to sustain the glow discharge. The power supplies are therefore small and easy-to-use devices.

Table 1.8 Visible lines of a HeNe laser.

Transition	Wavelength (nm)	Color	Rel. output power
$3s_2$–$2p_{10}$	543.5	Green	0.2
$3s_2$–$2p_8$	594.1	Yellow	0.4
$3s_2$–$2p_6$	611.8	Orange	0.5
$3s_2$–$2p_4$	632.8	Red	1.0

Also, HeNe lasers have smaller linewidths than diode lasers and, thus, longer coherence lengths. This is why they are still more suitable for interferometric applications than semiconductor lasers.

There are also lasers with liquid amplification media, so-called dye lasers. These are, however, quite large devices that have to be pumped with, for example, high-power argon-ion lasers. In general, their large setups and high power consumptions make them unsuitable for use in optical sensor systems.

As we have seen, lasers are usually light sources of high spatial and temporal coherence. By coupling highly intensive, pulsed laser light into, for example, optical fibers, nonlinear processes inside the material lead to the generation of a wide continuous spectrum of light. Reference [18] gives a good starting point for any reader who is interested in more details about the nature of these nonlinear effects. The result is a laser system of low temporal, but still large spatial coherence, a so-called supercontinuum light source. A commercial model covers a spectral range between the UV and infrared parts of the spectrum (395–2100 nm, [19]). The repetition rate is 80 MHz, and the output power reaches 700 mW. Potential applications cover chromatic-confocal sensors, white-light interferometers (see below) or multiwavelength fluorescence microscopy [18].

3
Photodetectors

Before any evaluation of the information that the received light carries can start, it has to be transformed into electrical signals. All processes in which light incident onto a surface produces a current or a voltage are therefore processes around which photodetectors can be designed. The most important of these processes is the photoeffect upon which the working principles of different types of detectors are based.

3.1
Photomultipliers

The working principle of the photomultiplier is based on the external photoelectric effect [20]. In 1905, Albert Einstein explained this effect by means of the photon hypothesis of light, and he was therefore awarded the Nobel Prize in Physics in 1921. When light hits a (metal) plate, photoelectrons are released from the surface of the material. Their kinetic energy, E_{kin}, is proportional to the frequency of the incident light, v, that is, its color. There is a minimum frequency, corresponding to a minimum energy, the so-called work function, W, that is required to release electrons from the surface. This is described by Einstein's equation

$$E_{kin} = h \cdot v - W \tag{1.11}$$

As an important consequence, the kinetic energy of the photoelectrons does not depend on light intensity. This means that if the light source does not emit photons with sufficient energy, there will be no photoelectrons leaving the surface, no matter how intense the light source, that is, the number of photons per unit area and time, is.

If, however, the energy of the photons is large enough, increasing light intensity will increase the number of photoelectrons and, thus, the photocurrent. As this current may be very low for a small number of photons, it can be amplified with an arrangement as depicted in Figure 1.15, a so-called photomultiplier tube (PMT): an evacuated glass tube contains a photocathode, an anode, and several additional electrodes, the "dynodes." The material of the photocathode determines the spectral sensitivity of the detector. The external resistor network divides the negative

Optical Sensors: Basics and Applications. Jörg Haus
© 2010 WILEY-VCH Verlag GmbH & Co. KGaA, Weinheim
ISBN: 978-3-527-40860-3

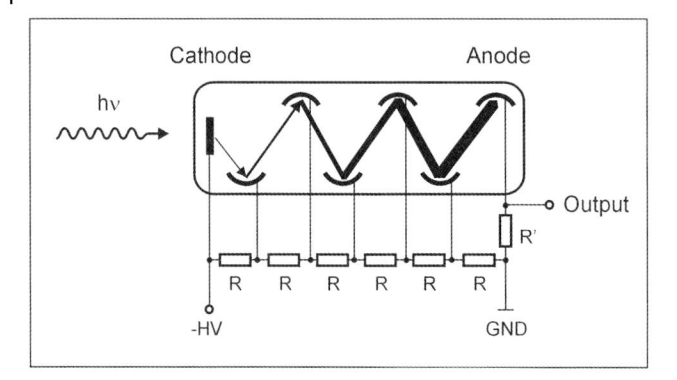

Figure 1.15 Photomultiplier with resistor network. (−) HV: (negative) high voltage, GND: ground, R, R′: resistors.

supply voltage up into equal potential drops between the dynodes. When a photoelectron hits the first dynode after being accelerated by the first potential drop, secondary electrons will be released. Each of these will hit the second dynode and, in turn, release additional secondary electrons. This avalanche effect will lead to a current amplification from stage to stage. The resulting anode current is transformed into a voltage across a resistor (R′ in Figure 1.15). This voltage is proportional to the light intensity incident on the photocathode.

Always depending on the work functions of the photocathode materials, PMTs have their maximum spectral sensitivities in the ultraviolet part of the electromagnetic spectrum. However, there are special materials, in particular multialkali cathodes (Sb–Na–K–Cs), with increased sensitivities in the visible and near-infrared [21].

PMTs are characterized by their wavelength-dependent quantum efficiency, $\eta(\lambda)$, the ratio between the number of released photoelectrons and the number of incident photons. When S_λ is the spectral sensitivity of the photocathode in A/W, and with the wavelength given in nm, the quantum efficiency can be calculated to [21]:

$$\eta(\lambda) = \frac{S_\lambda \cdot 1240}{\lambda} 100\% \tag{1.12}$$

As an example, a spectral sensitivity of 30 mA/W at 400 nm results in a quantum efficiency of about 10%. As the current amplification or gain, g, can reach 10^7 for a supply voltage of around 2 kV, PMTs must not be illuminated with high intensities. This would result in high-current discharges between the dynodes and to a destruction of the PMT.

A PMT will yield an output current even if it is kept in the dark. This dark current, I_d, is caused by, for example, thermionic emission from the electrodes, field emission, and by radioactivity from the environment, cosmic rays, or from the materials used in the PMT. This dark current is usually in the range of 1 pA and increases with increasing supply voltage. Its AC components determine the

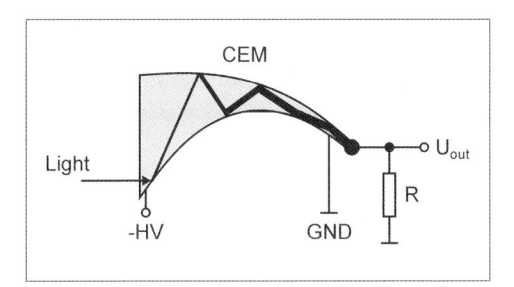

Figure 1.16 Channel electron multiplier (CEM). (−) HV: (negative) high voltage, R: resistor, U_{out}: output voltage.

detector noise current, I_{noise}, following Eq. (1.13) (e represents elementary charge and Δf: detection circuit bandwidth):

$$I_{noise} = \sqrt{2e\, I_d\, g\, \Delta f} \tag{1.13}$$

When a laser pulse arrives at the photocathode, there will be a corresponding voltage pulse at the PMT output. This output is characterized by the rise time, the time that the pulse takes to rise from 10% to 90% of its peak height, and by it fall time, the time that the voltage takes to drop back from 90% to 10% of the peak height. As the peak has an asymmetric profile, its full width, which is in the range of just a few nanoseconds, is about 2.5 times its rise time [21].

The circuit shown in Figure 1.15 is suitable for DC operation of a PMT. With a sufficiently fast external circuit, however, it is also possible to perform single-photon counting at extremely low intensities: each pulse will correspond to one incident photon, and its height will be a measure of its energy.

Photomultipliers exist in different designs, but with a mostly cylindrical tube. The light-receiving window may be head-on, but also side-on. PMT assemblies are generally larger than photodiodes, because apart from the tube itself, one needs a socket to hold the tube, the resistor network, and a high-voltage supply. However, miniaturized types exist that fill volumes of just a few cubic centimeters and come with an internal high-voltage circuit fed by a supply voltage of just a few volts.

While the photomultipliers described so far have discrete dynodes, there are also electron multipliers with continuous dynodes. They are usually realized in the form of tubes of highly resistive materials, for example, a particular lead silicate glass, covered with a semiconductor layer on their insides. This design does not only ensure the generation of secondary electrons on the tube walls, but also the voltage division between cathode and anode. These devices are single-channel electron multipliers (CEMs). A CEM usually has an inner diameter of about 1 mm and a wall thickness of between one and a few millimeters. Figure 1.16 shows a particular design with a funnel-shaped, curved tube with the cathode at the wider end and a collecting anode at its rear end. The charge impulses arriving at the anode are transferred into voltage pulses, for example, by means of a resistor.

One of the most common designs is the "Channeltron" [22]. As in the case of PMTs, the acceleration voltage is in the range of a few kilovolts. Analog gains can be as high as 10^7; pulse counting gains are larger by one order of magnitude. There are also two-dimensional arrays of cylindrical CEMs arranged and supplied in parallel, referred to as microchannel plates (MCPs), but they are mentioned here just for the sake of completeness.

3.2 Photodiodes

Photodiodes require no high voltage and no special detector housing so that they are far easier to handle and operate than photomultipliers. They are, however, inferior in their noise characteristics. Photodiodes utilize the effect of photons onto the charge carriers in the depletion zone in the pn-junction of a semiconductor diode: when photons are absorbed, they create electron–hole pairs. As the charge carriers remain inside the material, this is called the internal photoeffect. An applied reverse voltage forces the charge carriers to drift toward the external electrodes, and this produces a photocurrent that is proportional to light intensity. This mechanism can be understood as an inversion of the working principle of LEDs and diode lasers.

Some of the charge carriers that travel across the pn-junction will get lost due to recombination processes. Due to these recombinations, bandwidths of pn-diodes do not exceed values of about 10 MHz, but architectures that enable the charge carriers to travel faster across the junction provide higher reverse voltages. One of these architectures is a sequence of a p-doped, an undoped (intrinsic), and an n-doped semiconductor material, a so-called PIN diode. The electrical field across the intrinsic layer is constant and provides constant acceleration for the charge carriers. With these photodiodes, bandwidths in the GHz range have become possible. A further increase in the reverse voltage will amplify the photocurrent by two orders of magnitude because the charge carriers are accelerated to such an extent that they release additional charge carriers by ionizing collisions. These avalanche photodiodes (APDs) are semiconductor alternatives for PMTs in the detection of low intensities and reach bandwidths in the GHz range as well.

Apart from the architectures of the semiconductor material, also the sizes of the active areas determine the dynamic properties of photodiodes: the larger these areas are, the slower the response of the detector.

Semiconductor diodes have logarithmic current/voltage (I/U) characteristics. In photodiodes, however, there is an additional contribution from the photocurrent. This will be proportional to the number of incident photons, that is, to the incident radiant power, P_L. With the reverse current, I_S, with $U_T = kT/e$, and with the spectral sensitivity S_λ, it is

$$I = I_S \cdot (e^{U/U_T} - 1) - S_\lambda \cdot P_L \tag{1.14}$$

Equation (1.14) means that increasing light intensities shift the I/U curve down the current axis (Figure 1.17).

Photodiodes are typically made of silicon or germanium. The spectral sensitivity of Si reaches from about 300 nm to 1100 nm and peaks at around 950 nm, beyond the visible spectrum, whereas the maximum spectral sensitivity of Ge is at around

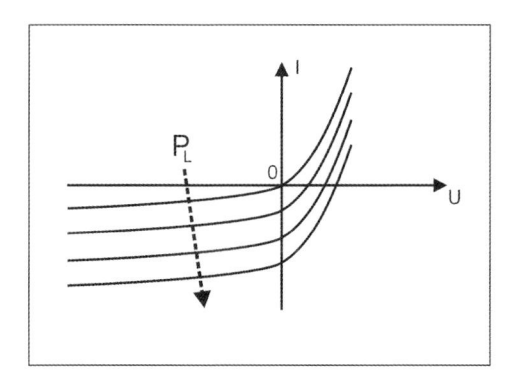

Figure 1.17 Current-to-voltage (I/U) characteristics of a photodiode in the photoamperic mode of operation.

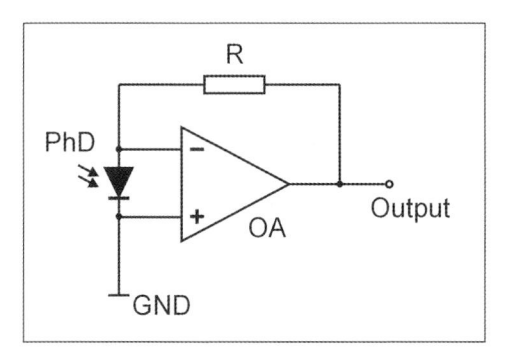

Figure 1.18 Photodiode: photoamperic mode. PhD: photodiode, R: resistor, GND: ground, OA: operational amplifier.

1450 nm and drops to zero at about 1800 nm. There are also semiconductor materials with increased sensitivities at larger wavelengths. One of these materials is indium-gallium-arsenide (InGaAs) with a spectral sensitivity range from about 900 nm to about 2500 µm, with its maximum at about 1650 nm wavelength. At the peak of their sensitivity curves, Si photodiodes have spectral sensitivities of about 0.5 A/W, while Ge photodiodes can reach values of 0.9 A/W and InGaAs photodiodes values of more than 1.0 A/W.

A photodiode is usually integrated as a current source into an electronic circuit. To this end, it is connected directly to the input terminals of a current-to-voltage converter (Figure 1.18). In this photoamperic mode of operation, the output voltage is proportional to the photocurrent and, thus, to light intensity. This preamplifier setup can either be realized with single components, or with integrated photodiode/preamplifier circuits contained in one single optoelectronic element.

3.3
Other Detector Types

Photoresistors are devices based on the internal photoeffect in semiconductor crystals. Discovered in 1873 with selenium, the effect is, for example, utilized in the exposure meters of photocameras. At room temperature, their CdS detectors

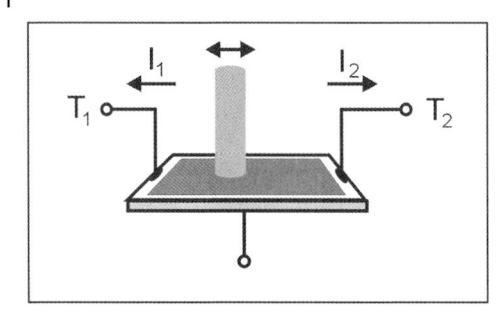

Figure 1.19 Position-sensitive device (PSD). T_1, T_2: terminals, I_1, I_2: currents.

have maximum spectral sensitivities at $\lambda_{max} = 500$ nm. Other examples for detector materials are PbS ($\lambda_{max} = 2.4\,\mu m$) and HgCdTe ($\lambda_{max} = 10.6\,\mu m$).

Photodiodes with at least two terminals are capable of detecting the positions of light spots on their active surfaces and are therefore denoted as position-sensitive devices (PSDs). A one-dimensional PSD has three terminals, one at its center and one on either side. It detects the light spot position between the two outer terminals (Figure 1.19).

Divided by the total photocurrent, $I_1 + I_2$, the difference between the photocurrents I_1 and I_2 is directly proportional to the light spot position, P, on the PSD with length L according to

$$P = \frac{L}{2} \cdot \frac{I_1 - I_2}{I_1 + I_2} \tag{1.15}$$

There are also two-dimensional PSDs with five connectors that indicate the two-dimensional position of the light spot in analogy to Eq. (1.15). One-dimensional PSDs typically come with light-sensitive areas of about 10 mm × 2 mm, and they reach linearities of around 0.1%. They are employed in distance, position, and angle measurement systems based on optical triangulation. Similar to PSDs, segmented photodiodes have several independent active areas on one chip. With a suitable wiring, the light spot positions are also calculated according to Eq. (1.15).

Infrared light can indirectly be detected by its thermal effects. The basic effect in bolometers is the temperature dependence of the electric conductivity: in metals, the conductivity decreases with increasing temperature, while in semiconductors, it increases. One important element is the Pt100 resistor for the measurement of temperatures between $-100\,°C$ and $+400\,°C$. The effect is usually exploited by integrating the resistor into a Wheatstone bridge.

Thermoelements are based on the Seebeck effect and are simply made up of two wires soldered together on both ends, with one junction held at a defined temperature as reference, for example, $0\,°C$. The voltage difference between the two junctions is then a direct measure for the temperature of the second junction. The effect can be multiplied by connecting several thermoelements in series to a so-called thermopile detector. A typical thermopile detector has a sensitivity of 20 V/W between 5 μm and 14 μm wavelength [23]. Thermopiles are used, for example, in power meters for high-power lasers and in noncontact infrared thermometers.

3.4 Imaging Detectors

The detectors described in the last chapters are only capable of recording one single channel. If, however, some spatial resolution is required, or in other words, if a one- or two-dimensional image of a scene shall be captured, one needs a one- or two-dimensional detector array, respectively.

One-dimensional detector arrays are used for so-called line cameras. Basically, they consist of a light-sensitive area and a readout zone. The light-sensitive area is composed of single picture elements (pixels) a few micrometers in size, with square or rectangular shape. The smaller the pixels, the higher their resolution, but the less light is collected on the active area, and the higher the noise level becomes. In the pn-junctions of these pixels, charge carriers are generated by incident light. In contrast to photodiodes, however, these charge carriers are collected so that their number is proportional to light intensity. The readout zone basically consists of a shifting register that shifts the single charge packages sequentially from the pixels to a readout amplifier. The output of this charge-coupled device (CCD) is an intensity-proportional voltage. As the image is recorded in parallel, but the output is not, several ways exist how the image is put-together with two-dimensional CCD arrays. Most video cameras employ interlaced CCDs that transfer only half-images, that is, images that contain only the even and the odd line numbers, respectively. Progressive-scan CCDs, on the other hand, transfer one line after the other (Figure 1.20). For sensors with several megapixels, the readout frequencies, that is, the number of pixels per second, can be as high as some 10 MHz for frame rates of some 10 frames per second.

For very small light intensities, as they may have to be captured in fluorescence microscopy, cooled cameras with special CCD designs have become available. These electron-multiplying CCDs (EMCCDs) are mostly designed like standard CCDs, but they feature an additional, multiplying register that follows the shifting register. Each cell of this multiplying register is supplied with a high voltage so that multiplication of the shifted charge occurs by way of impact ionization. EMCCDs can detect wavelengths between 250 and 1000 nm with quantum efficiencies of more than 90% and are thus capable of detecting single-photon events. Furthermore, the readout rates can reach 35 MHz so that frame rates of 500 frames per second become possible [24].

In contrast to CCDs, complementary-metal-oxide-semiconductor (CMOS) sensors feature current-to-voltage converters and amplifiers in every single pixel. Each pixel is read out without the need for shifting its signal as in a CCD; the additional circuitry, however, requires extra space. The fill factor, that is, the ratio between the total light-sensitive area to the total area of the chip, is thus reduced. This also implies that the overall light sensitivity of CMOS sensors is smaller than

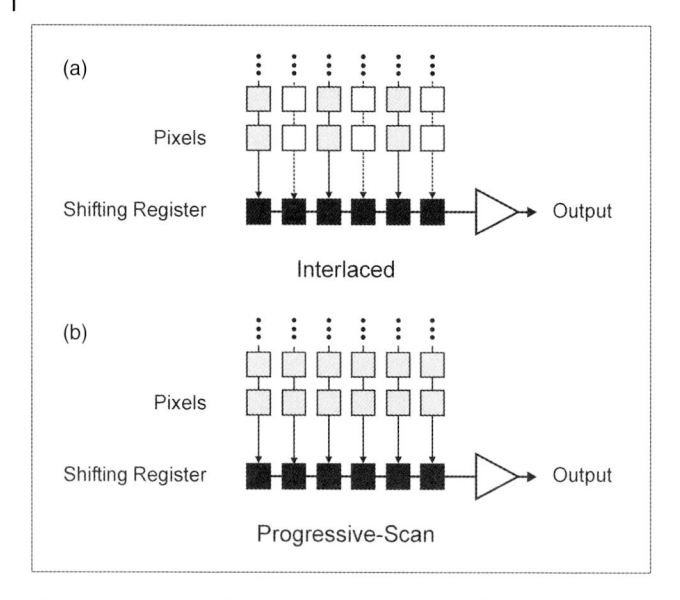

Figure 1.20 CCD readout techniques. (a) Interlaced, odd and even lines are read out separately. (b) Progressive-scan, the lines are read out one after the other.

that of CCDs. Further miniaturization and microlenses on every pixel have helped to overcome these drawbacks. On the other hand, the power consumption of CMOS sensors is smaller than that of CCDs.

CMOS sensors are widely used as image sensors in digital cameras, from single-lens reflex (SLR) cameras to mobile phone cameras. In other spectral regions, they are less sensitive than CCDs, which generally renders them unsuitable for surveillance cameras that require high sensitivity in the infrared. For both sensor types, resolution increases with increasing number of pixels. As the sizes of the sensors do not change significantly, this always comes with decreasing pixel sizes. This, however, means that for a given image, the intensity per pixel decreases as well so that the signal-to-noise ratio becomes smaller. Of course, more megapixels always mean higher resolution, but also a generally more adverse noise behavior. Also, with high camera resolutions, one should always try to estimate if the camera optics is indeed capable of providing the same optical resolution. If not, a high camera resolution would be completely useless. This is often the case with digital cameras in microscopy: here, the optical resolution of the objective is determined by its numerical aperture (see Section 6.4). A simple calculation shows that a camera with 1.3 megapixels and a ½ inch detector chip is sufficient for all detail that an objective with a numerical aperture of 0.65 can resolve.

Neither the pixels of a CCD sensor nor those of a CMOS sensor are color-sensitive, their output is proportional to light intensity only. The most direct, but also most expensive solution is to direct the image onto three identical sensor arrays, each equipped with a single filter for one primary color (blue, green, red). This will, however, come with a comparably complicated optical setup. An alternative

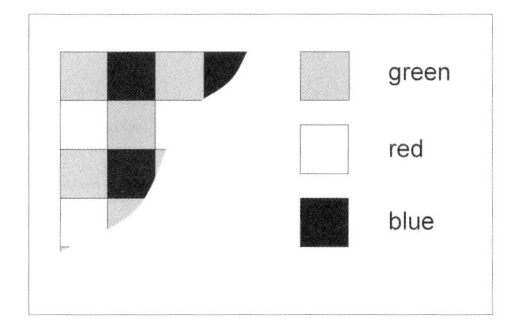

Figure 1.21 A Bayer filter.

is the so-called Bayer filter (Figure 1.21), an array of filters that transmit the primary colors. Along the odd and even lines, there is a periodic sequence of green/blue and green/red filters, respectively. As a consequence, there are twice as many green filters as blue and red ones, which models the higher sensitivity of the human eye in the green part of the spectrum. The "true" color of a pixel is thus calculated by interpolation of different neighboring color channels. A recent innovation with the aim to yield a direct image with blue, green and red color components at full resolution is the Foveon X3 sensor. It incorporates three image sensors stacked on top of each other, realized through the manufacturing process of the sensor array. Photons of different wavelengths have different penetration depths into the semiconductor material so that the blue ones are captured near the surface of the chip, the green ones in the central, and the red ones in the lowermost sensor array. Without the need to interpolate the color pixels, the images appear sharper than that with a Bayer filter [25]. The first digital SLR camera with this chip was introduced in the market in 2003, but the majority of the cameras on the market still work by the conventional technology.

3.5
Detector Noise

In general, noise is every nondeterministic contribution to a detector signal that reduces its dynamic range. Due to their statistical nature in the time domain, the characteristics of noise can better be examined in the frequency domain. The noise spectrum is composed of two types of noise: below frequencies f of about 100 Hz it is dominated by the so-called flicker noise which follows a $1/f$ law. Its origin has not yet been fully understood, but generation/regeneration processes in the pn-junctions are one possible source. For frequencies from 100 Hz up to 100 THz, there is only frequency-independent thermal and "shot" or "white" noise. One of the sources for shot noise is the noise of the photodiode dark current. The second source is photon noise which is proportional to the square root of the light intensity. The rms value of the shot noise current, I_{rms}, only depends on the average current, I_0, and the bandwidth, Δf, and it follows from Poisson's statistics:

3 Photodetectors

Table 1.9 Comparison of different photodetector types: Si PIN photodiode, photomultiplier, APD.

Parameter	Si PIN photodiode	Photomultiplier	APD
Spectral range (nm)	320–1100	185–650	400–1000
Active area (mm^2)	25	~40[a]	~7
λ_{max} (nm)	960	340	905
$S_\lambda(\lambda_{max})$[h] (A/W)	0.72	$3.6 \cdot 10^{5\text{b})}$	60[c]
Supply voltage (V)	–	1000	–
Breakdown voltage (V)	–	–	300
Typical dark current (nA)	0.4[d]	0.5[e]	1[c]
Noise current (pA/Hz^{½})	0.02[d)f)]	–	0.5
Dark current voltage (mV)	–	<0.1[g]	–
Rise time (ns)	50[d]	1.4	0.5[c]

a) Photocathode area.
b) Anode.
c) Gain $M = 100$.
d) Reverse voltage $V_R = 10\,\text{V}$.
e) After 30 min.
f) Calculated from NEP and $S_\lambda(\lambda_{max})$.
g) As part of module H8569.
h) S_λ, spectral sensitivity; λ_{max}, wavelength with maximum spectral sensitivity.

$$I_{rms} = \sqrt{2\,e\,I_0\,\Delta f} \tag{1.16}$$

Detector noise is also characterized by the incident radiant power for which the signal-to-noise (S/N) equals 1. This particular radiant power is denoted by the noise-equivalent power, NEP, and its unit is W/Hz$^{1/2}$:

$$\text{NEP} = \frac{\text{Noise Current}\,(A/\sqrt{\text{Hz}})}{\text{Spectral Sensitivity}\,(A/W)} \tag{1.17}$$

A typical value for the NEP of a photodiode is $10^{-14}\,\text{W/Hz}^{1/2}$. For a PMT, it is two orders of magnitude smaller. The inverse of the NEP is the detectivity, D. As the noise of a photodiode is proportional to the square root of its light-sensitive area, A, the specific detectivity, D^*, is defined as

$$D^* = D \cdot \sqrt{A} = \frac{\sqrt{A}}{\text{NEP}} \tag{1.18}$$

The unit of D^* is Hz$^{1/2}$W^{-1}cm. Flicker noise causes the specific detectivity to decrease for low frequencies. For high frequencies, on the other hand, it will also be small due to the decreasing sensitivity of the detector.

As a summary of this chapter, Table 1.9 shows the specifications of three types of photodetectors: a Si PIN photodiode [26], a PMT [27], and an APD [28].

4
Optical Elements

There is a large toolbox of optical elements that can be incorporated into an optical sensor. They range from mirrors and lenses over dispersive elements like prisms and gratings to filters, optical fibers and optical modulators. Optical elements are required to guide light, to form and transform images, to manipulate beam shapes, to limit and select wavelength ranges, and to modulate light. Detailed descriptions of these elements will easily fill complete textbooks, and readers interested in more details are referred to Refs. [1] and [29]. This section will start, however, with a short compilation of the most important optical properties of optical materials available today.

4.1
Optical Materials

The optical properties of a material can be completely described by its complex, wavelength-dependent index of refraction, $n'(\lambda)$:

$$n'(\lambda) = n(\lambda) + i\kappa(\lambda) \tag{1.19}$$

Here, $i^2 = 1$. The real part, $n(\lambda)$, determines how light rays are refracted when they cross an interface between two materials. Its wavelength dependence leads to the effect of dispersion, and Section 4.3 gives an overview about how this effect is exploited by so-called dispersive optical elements. Optical materials like glass or quartz usually show what is called "normal" dispersion, which means that blue light has a higher refractive index than red light, and its angle of refraction is larger:

$$\frac{dn}{d\lambda} < 0 \tag{1.20}$$

The imaginary part of the refractive index, $\kappa(\lambda)$, is the extinction coefficient and describes absorption losses inside the material. Its wavelength dependence is represented by the material's absorption spectrum. Optical glass usually does not transmit ultraviolet light with wavelengths below about 320 nm. Figure 1.22 shows the typical internal transmittance of an optical glass [30]. This internal transmit-

Optical Sensors: Basics and Applications. Jörg Haus
© 2010 WILEY-VCH Verlag GmbH & Co. KGaA, Weinheim
ISBN: 978-3-527-40860-3

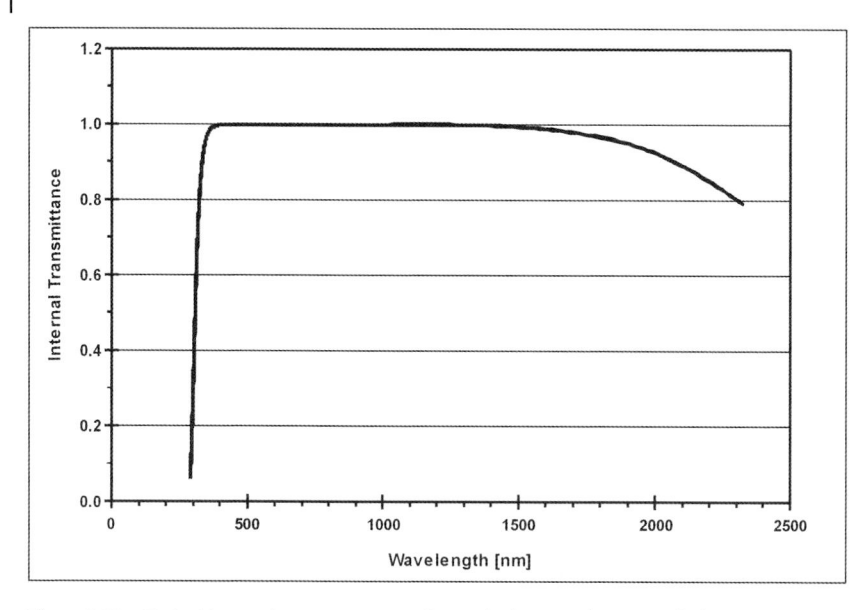

Figure 1.22 Typical internal transmittance (10 mm thickness) of an optical glass.

tance is the percentage of the intensity that passes the material, without the reflective losses at its surfaces. As can be seen from the chart, this type of glass is perfectly transparent over the entire visible spectrum, from about 400 nm to about 800 nm, and also far into the infrared spectrum.

Higher transmissions in the ultraviolet spectrum can only be achieved with quartz glass. Figure 1.23 shows the typical internal transmittance [30] of a glass used in ultraviolet optics, for example, for lithographic applications. Here, transmittance already sets in at around 200 nm wavelength.

Besides the different types of optical glass, optical elements can also be manufactured from polycarbonates, a type of transparent plastic. One particular material [31] has a transmittance of around 90% at a thickness of 2 mm and reaches a refractive index of more than 1.57 over the entire visible spectrum. The ultraviolet cutoff is at a little less than 400 nm, depending on the type of material, and there are also absorption bands in the near infrared. The combination of small weight and high refractive index makes polycarbonate a preferred material for correction glasses. In contrast to glass, it is very resistant to impacts, but must be handled carefully to avoid scratches. As polycarbonates can be injection molded, they can very conveniently be employed to produce illumination optics for all types of light sources, for example, in the automotive industry. It should be noted, however, that their properties are not yet sufficient to manufacture high-quality imaging optics – for these, designers still make use of optical glass.

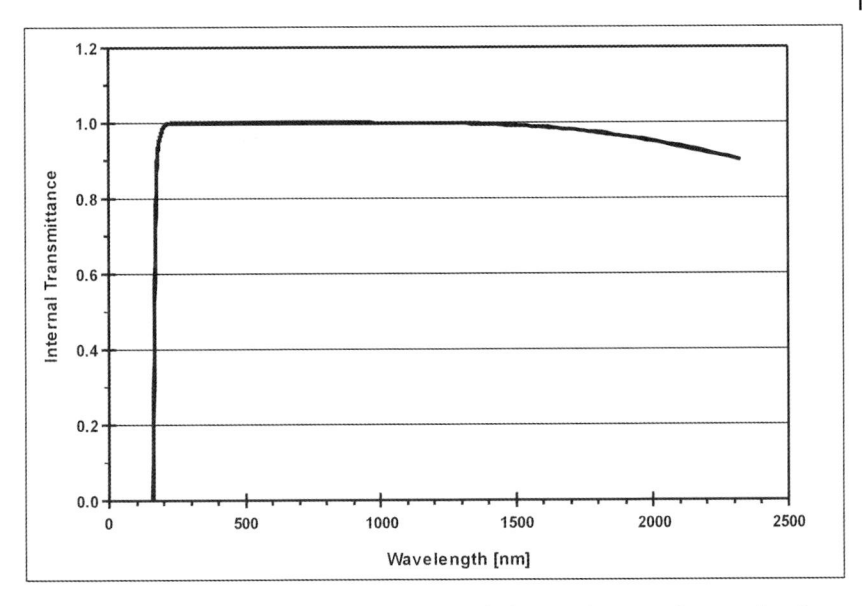

Figure 1.23 Typical internal transmittance (10 mm thickness) of an optical quartz glass for ultraviolet optics.

4.2
Mirrors, Prisms and Lenses

With a mirror, light can be directed quite easily. It works by the law of reflection, according to which the angle of reflection equals the angle of incidence. In addition, incident ray, reflected ray, and the surface normal are in the same plane. Mirrors are manufactured by either polishing metal surfaces, or by coating, for example, glass substrates. Silver and aluminum, for example, have a reflectance of more than 95% over the entire visible range of the spectrum, but become transparent for wavelengths shorter than 300 nm. Mirrors made of gold or copper have smaller reflectances in the short-wavelength range, and therefore appear yellow-red. Even higher reflectances are required for optical setups in which multiple reflections occur. These would cause a considerable decrease in light intensity, for example, in laser resonators. As an example, six reflections on aluminum mirrors with reflectances of 95% will reduce the incident intensity to $0.95^6 = 0.66$. The mirrors employed in these applications are comparably sophisticated coating structures on glass substrates with reflectances of more than 99%. They are quite expensive and the coating layers are sometimes quite soft and have to be handled with extreme care.

Mirrors can be flat or curved. Spherical or parabolic mirrors are used as reflectors in illumination systems, but also as elements in imaging elements, for example, in Newton-type telescopes. These mirrors need to be of high surface

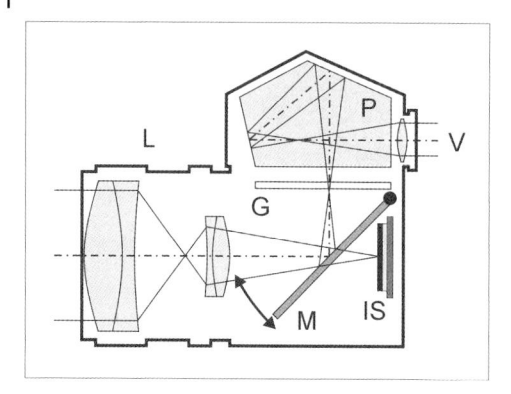

Figure 1.24 Schematic optical setup of a single-lens reflex camera with pentaprism viewfinder. L: lens, M: flip mirror, IS: imaging sensor, G: ground glass screen, P: pentaprism, V: viewfinder.

qualities so that the incoming light wavefront remains as undisturbed as possible. In contrast to lenses (see below), mirrors show no chromatic aberrations, which means that the angles of reflection are identical for all incident wavelengths.

One disadvantage of mirrors is the need for sometimes bulky and complex mounts and adjustments. Exploiting the total reflexion of light on interfaces between optical materials of different refractive indices (see also Section 4.6), changing the direction of a light beam can also be achieved with prisms. One example for the employment of mirrors and prisms is the optical setup of SLR cameras: the camera lens projects the image onto the imaging sensor. A portion of this light is reflected out of the beam path by a flip mirror that directs the image to the camera's viewfinder. This mirror cannot be made from a prism because this would make it too heavy for the required dynamics. Here, the photographer does not only observe an upright image, but also one with no parallactic displacement with regard to the final image. Figure 1.24 shows the system schematically and how the pentaprism reflects the image twice. What this image does not depict is that the top of the pentaprism is formed like a roof, which was represented by the typical designs of (analog) SLR housings until a couple of years ago. This roof design is necessary to reverse the left and right sides of the image to produce not only an upright, but also a laterally correct image.

As digital SLRs become more and more popular with their prices decreasing dramatically, the pentaprism is often replaced by a pentamirror. It is cheaper to manufacture, but it does not show the complete image. Also, it may not be as bright as a pentraprism viewfinder.

One disadvantage of prisms is their heavy weight and, thus, comparably high price. Also, when prisms are to be employed in laser applications with large output powers, any air inclusions must be avoided in the manufacturing process. Otherwise, these inclusions would lead to locally increased absorption that could cause thermal damage to the prism.

The effect of lenses is based on Snellius' law of refraction. The focal length, f, of a thin spherical lens with refractive index n and with the front and back radii R_1 and R_2 is

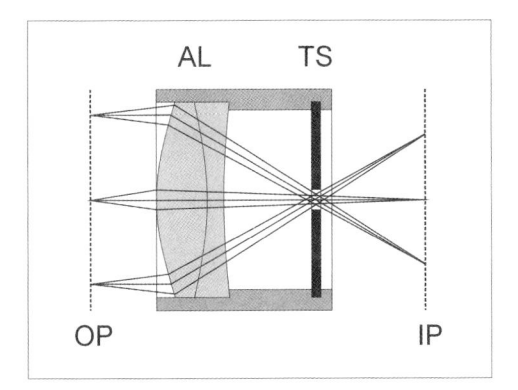

Figure 1.25 A telecentric lens. OP: object plane, IP: image plane, AL: achromatic lens, TS: telecentric stop, placed in the back focal plan of AL.

$$\frac{1}{f} = (n-1)\left(\frac{1}{R_1} - \frac{1}{R_1}\right) \tag{1.21}$$

When the spherical surface of the lens is convex toward the beam, the radius is positive, while it is negative, when the surface is concave. One example is a biconvex lens with two equal radii $R_1 = R = -R_2$, for which f becomes positive (a positive lens). For a biconcave lens, it follows that $f < 0$, which is therefore called a negative lens.

A single lens shows several types of aberrations. The wavelength-dependence of the refractive index in Eq. (1.21) results in wavelength-dependent focal lengths. The corresponding images show colored fringes instead of sharp contours. A so-called achromat is a combination of a positive and a negative lens, and their respective refractive indices are balanced to yield equal focal lengths for red and blue light. The remaining chromatic error, the so-called secondary spectrum, can be corrected by adding a third lens with a new appropriate combination of the refractive indices of all three lenses. These apochromatic lenses are used in all applications in which true-color imaging is demanded, for example, in high-end microscope objectives or in astronomical telescopes.

Other aberrations influence the geometry of the image. Distortions are the variations of the optical magnification from the center of an image to its edges. Pincushion distortion describes the increase, and barrel distortion the decrease in magnification with increasing distance from the center. The images of simple achromats are generally not "flat," which means that when the center of the image is in focus, its periphery is not. Image flatness can be increased by sophisticated optical designs, and they result in plan (-achromatic) objectives. In general, the higher the state of correction of an optical system, the larger the number of optical elements, and the higher its price.

One interesting example for an optical system design shall be given in Figure 1.25. It shows an objective which is "telecentric" on its object side. This kind of objective is often used in image processing applications when perspective distortions must be avoided, for example, when dimensions have to be measured. The objective has a telecentric stop in its back focal plane so that only rays parallel (or with a small angle) to the optical axis are used to form the image. As no ray with a large angle of incidence can reach the image plane, no perspective distortions can occur. As a necessary consequence, the front lens must be as large in diameter as the object, and quite a lot of light is lost at the telecentric stop.

For readers interested in a detailed description of the theory of optical imaging, Ref. [1] is a good starting point.

4.3
Dispersive Elements: Prisms and Gratings

As described above, dispersion is the dependence of the refractive index on the light wavelength. This is the effect behind the optical prism that splits white light up into its spectral colors. The central property to characterize the dispersion of a dispersive optical element is its spectral resolving power, $\lambda/\Delta\lambda$. According to the Rayleigh resolution criterion, two spectral lines cannot be discriminated unless their spacing is larger than their widths. For a prism with a base length S, the spectral resolving power is:

$$\frac{\lambda}{d\lambda} = S \cdot \frac{dn}{d\lambda} \tag{1.22}$$

The dispersion in the visible spectrum amounts to about $1000\,\text{cm}^{-1}$, which corresponds to a change in the refractive index of about 10^{-4} when the wavelength changes by 1 nm. Following Eq. (1.22), a prism with a base length of 5 cm has a spectral resolving power of about 5000, and this means that it can resolve the two yellow sodium D lines at about 589 nm wavelength: as their line spacing is $\Delta\lambda = 0.6$ nm, $\lambda/\Delta\lambda$ is about 1000.

A second important dispersive element is the diffraction grating. Here, the decisive parameter is the grating constant, a, the spacing between two adjacent grooves. The far-field intensity distribution (Fraunhofer diffraction) of diffracted monochromatic light with an angle of incidence, θ_i, onto a reflective grating shows intensity maxima in all directions θ_d for which

$$a(\sin\theta_d - \sin\theta_i) = m \cdot \lambda \quad m = 0, \pm 1, \pm 2, \pm 3, \ldots \tag{1.23}$$

When using polychromatic light, its different wavelengths in one diffraction order, m, become diffracted into different angles. According to Eq (1.23), and in contrast to a prism, the diffraction angle increases with increasing wavelength. From Eq. (1.23), one finds an expression for the angular dispersion, $d\theta/d\lambda$:

$$\frac{d\theta}{d\lambda} = \frac{m}{a\cos\theta_d} \tag{1.24}$$

In the 0th diffraction order, that is, for $m = 0$, no dispersion occurs, and all wavelengths are found in the same spot. For white light, this diffraction order appears white again. The spectral resolving power of a grating increases with the diffraction order in which the spectrum is observed, and with the number of illuminated grooves, N:

$$\frac{\lambda}{d\lambda} = N \cdot m \tag{1.25}$$

For high resolving powers, measurements should thus be carried out in high diffraction orders, and with a light beam that is expanded as far as possible so that as many grating lines as possible contribute to the analysis. One particular grating is the so-called blazed grating. For one specific wavelength, the blaze wavelength, the diffraction angle for the first diffraction order ($m = 1$) equals the angle of reflection. Further information on optical gratings and their applications can be found in [32].

4.4
Optical Filters

Optical filters are used whenever the bandwidths of light sources must be limited or when light intensities shall be reduced without influence on the color temperature. When only a small spectral bandwidth is required, a line source is not necessarily a good solution in any case: the emission lines may be very weak. This is why continuum sources with appropriate filters are a practical alternative.

There are two different types of filters available: color-glass filters are, as their name indicates, made of ion-doped glass. The kind of doping determines the filter's transmission spectrum, and its thickness, d, the transmitted intensity, $I(d)$, according to Lambert–Beer's law

$$I(d) = I_0 \cdot \exp(-\alpha(\lambda) \cdot d) \tag{1.26}$$

with I_0 as the incident intensity and $\alpha(\lambda)$ as the wavelength-dependent absorption coefficient. Filters are characterized by the form of their transmission spectrum. Short- (long-) pass filters transmit only wavelengths shorter (longer) than a certain wavelength limit, bandpass filters transmit only over a narrow wavelength range, and neutral density filters show almost constant transmission over the visible range (Figure 1.26). Neutral density filters are also known as "gray" filters and they particularly serve to introduce a defined extinction to a continuum light source without changing its emission spectrum.

The bandwidths of these filters are, however, quite large. Also, the edges in the transmission spectrum are not very steep. Steep edges, however, may be required to filter closely spaced wavelengths. Small bandwidths and almost vertical edges in the transmission spectrum are achieved with interference filters. These filters consist of optical resonators that are made up of dielectric layer systems, that is, coatings on glass or quartz substrates. These layers are designed to yield the desired transmission spectrum by way of multiple optical interference. In this way, for example, a long-pass filter may block one of two wavelengths that are only about 1 nm apart. One disadvantage of interference filters is the dependence of their transmission spectrum on the angle of incidence of the incoming light. However, this can be exploited for tuning an interference filter: by tilting it slightly, the effective filter thickness changes, and the transmission spectrum of the filter

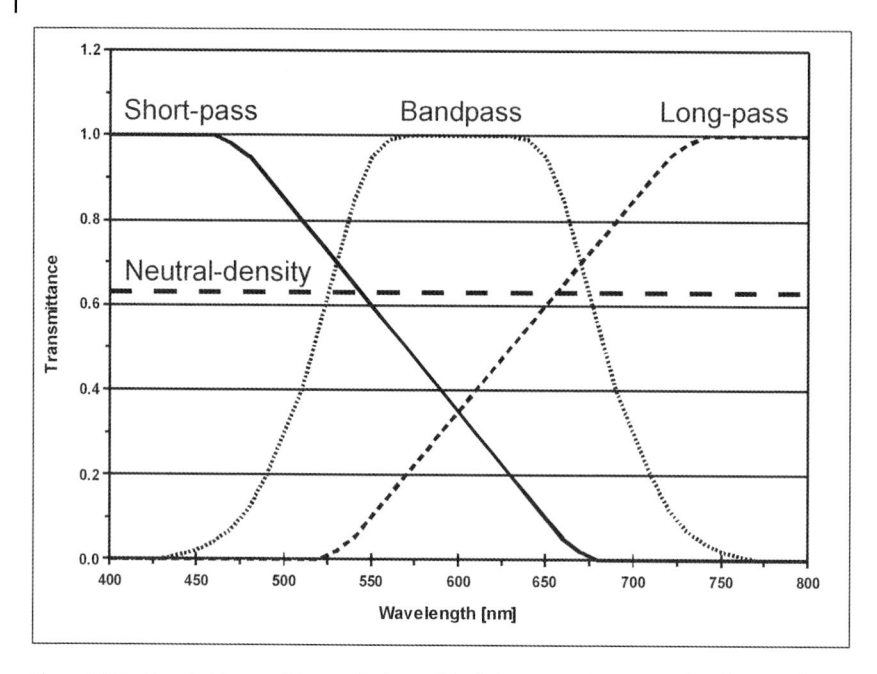

Figure 1.26 Spectral transmittances (schematic) of short-pass, long-pass, bandpass and neutral-density filters.

shifts. In general, coatings can be applied to influence the absorptive, transmissive and reflective properties of an optical surface. Moreover, they can be adjusted to optimize the resistance against environmental factors and wear and tear, or to show easy-to-clean or even self-cleaning effects. An overview over the state-of-the-art in optical coatings can be found in Ref. [33].

Dispersive elements can also be used as filters. In monochromators, light enters through the entrance slit, which is imaged via a prism or an optical grating onto the exit slit. Through the exit slit, only a small portion of the resulting spectrum passes, and its width determines the linewidth of the spectrum that leaves the monochromator. For a white-light source, arbitrary wavelengths can be chosen by a rotation of the dispersive element. When the exit slit is replaced by a photodetector array, several wavelengths can be monitored in parallel. These polychromators are, for example, employed in atomic spectrometry to detect the presence of several elements simultaneously.

4.5
Polarizers

The polarization of a light wave describes how its electrical field vector oscillates. Linearly polarized light oscillates in one plane, and the field vector of circularly polarized light rotates on a spiral as the light wave propagates. Except for lasers, the light from all the light sources described in Section 2.1 is unpolarized. Light can be polarized in four different ways, all of which are based on certain forms of optical anisotropy:

Light can be polarized by reflection and scattering. In photography, the former is widely used to eliminate disturbing reflections, and the latter enhances the contrast between sky and clouds in images; both effects are not too practical for the realization of polarizers. Particularly in the case of reflections, mirrors would have to be tilted by the so-called Brewster angle, and this kind of polarizer would require a lot of room in an optical setup. As to scattering, to the author's knowledge, no embodiment exists that makes use of this effect for the realization of polarization optics.

Dichroism, also denoted anisotropical absorption, is the dependence of absorption on polarization. This is the principle on which foil polarizers work [34]: When a foil of polyvinyl alcohol with embedded needle-shaped dye molecules is stretched into one direction, these molecules become aligned along the molecular chains and will thus absorb anisotropically. These foil polarizers are quite cheap, but they suffer from their poor closed-to-open-pair extinction ratio and their low open-pair transmission.

The fourth polarization effect is birefringence, or anisotropical refraction. It is a property of certain crystals, for example, calcite ($CaCO_3$), quartz, or magnesium fluorite (MgF_2). Their crystal structure shows one symmetry axis, the so-called optical axis. Light that travels under some angle to this axis is split up into two rays: the ordinary one is polarized perpendicular to the plane defined by the optical axis and the direction into which the light propagates. The extraordinary ray is polarized parallel to this plane. The indices of refraction for ordinary (n_o) and extraordinary (n_e) ray differ and lead to different angles of refraction. As an example, for calcite, these indices are $n_o = 1.65835$ and $n_e = 1.48640$ for $\lambda = 589.3$ nm [34]. Figure 1.27 shows well-known examples of highly efficient polarizers. They are composed of two crystals whose faces are cut under certain angles relative to the optical axis. The gap between these two components is filled with air or with a cement that matches the refractive indices. Birefringence results in the incoming ray being split up into two rays of which only one can exit the polarizer, and the other one is totally reflected at the gap and absorbed in one of the blackened side faces. This results in a highly polarized light output. If, however, the crystal's design allows both rays to leave the end face, this optical element is a polarizing beamsplitter, and both output rays will have the same intensity, which is approximately one half of the intensity of the incoming light. The Wollaston prism in Figure 1.27 is only one possible example for such a beamsplitter. The most common embodiment is a beamsplitter cube, made of two right-angle prisms with a dielectric layer in between.

As these types of polarizers and polarization beamsplitters introduce polarization only through birefringence and not through dichroitic absorption, almost no light intensity is lost, and the polarizing efficiency is very high – but not for all wavelengths. The short-wavelength limit for calcite is at around 213 nm, and if shorter wavelengths are employed, prisms made of quartz or MgF_2 must be used.

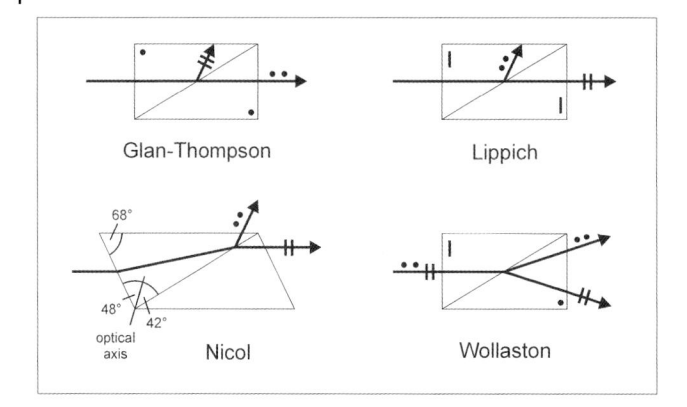

Figure 1.27 Polarizers of different types and the Wollaston prism as polarizing beamsplitter. Crystals: optical axis in/perpendicular to the paper plane (dashes/dots). Light rays: electrical field vector in/perpendicular to the paper plane (dashes/dots).

On the other hand, however, these optical materials are very soft and thus have to be manufactured and handled with extreme care. This and the generally higher material costs make these polarizing prisms far more expensive than foil polarizers.

4.6
Optical Fibers

Optical fibers are very convenient elements for guiding light from a light source to some distant target. They are not only employed in high-speed data transmission networks, but also for illumination purposes, particularly when there is a central light source the light of which shall be distributed to different distant positions.

The advantages of optical fibers is their small dimensions and weight, their flexibility, the absence of electromagnetic emissions, their insensitivity against electromagnetic interferences, and thus the ability to guide signals in electrically and magnetically disturbed environments. Optical fibers support high carrier frequencies and high signal bandwidths. Despite these advantages, light also suffers from damping effects during the transport, and when signals must be transported over large distances, repeaters have to be installed.

A first crucial step is to couple light into a fiber. There are special fiber couplers available which are, on principle, lenses with small focal lengths. As an alternative, particularly in experimental setups, microscope objectives with suitable numerical apertures can be used. The optical fiber itself consists of an inner core, which is surrounded by a cladding with an index of refraction that is always smaller than that of the core. Thus, total inner reflection on the core/cladding interface holds the light inside the core. Every path that a light beam can follow through the fiber

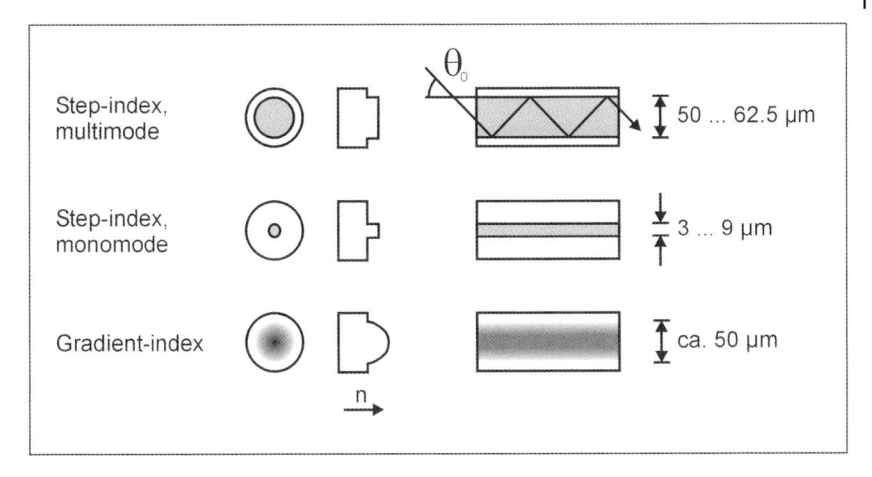

Figure 1.28 Types of optical fibers: step-index multimode, step-index monomode and gradient-index multimode. Left column: cross sections, right column: longitudinal sections. θ_0: acceptance angle.

is called a "mode." The more modes can propagate, the wider the distribution of possible path lengths becomes, and the signal phases are widely distributed. This so-called multimode dispersion leads to a smeared output signal. For longer transmission distances, the mode distribution must therefore be limited by employing monomode fibers, in which only a narrow distribution of light paths is possible.

There are three major types of fibers, classified by their radial refractive index distributions. Step-index profile fibers exist as mono- and multimode fibers (Figure 1.28). They have uniform, but different refractive indices in core and cladding. Between core and cladding, there is a steep decrease in the index of refraction of around 1%. While the outer diameter of the fiber is 125 μm, the core diameter is between about 50 and 60 μm for multimode fibers and between 3 and 9 μm for single-mode fibers.

As indicated in Figure 1.28, light can travel along a large variety of paths in multimode fibers. If the mode has a small angle to the optical axis, it is a mode of low order; if the angle is larger, it is a mode of higher order. Modes that enter the cladding may be reflected on the outer surface of the fiber, and these cladding modes can even propagate over distances of some hundreds of meters, in spite of their large damping constants.

Monomode fibers have a relative refractive index difference of around 0.1%. With a as the core radius, and n_1 and n_2 as the refractive indices of core and cladding, respectively, the normal fiber frequency, f_n, is defined as

$$f_n = \sqrt{\frac{\pi}{\lambda} a \frac{n_1^2 - n_2^2}{n_2}} \tag{1.27}$$

With this frequency, a monomode fiber is qualified by the following condition [34]:

Table 1.10 Acceptance angles (θ_0) and numerical apertures (NA) of different types of optical fibers.

Fiber material	θ_0 (°)	NA[a]	Approx. transmission range (nm)
Glass	26.5	0.45	400–1600
	32.5	0.54	
	35.0	0.57	
	40.0	0.64	
	60.0	0.86	
Quartz	12.5	0.22	250–1000
	15.0	0.26	
PMMA[b]	27.5	0.46	400–800
	30.0	0.50	

a) NA values valid for wavelengths below 587 nm, refractive index $n_0 = 1$.
b) PMMA, polymethyl methacrylate (acrylic glass).

$$1.8 < f_n < 2.4 \tag{1.28}$$

In gradient index fibers, the relative refractive index difference between core and cladding is between 1% and 2%. The index profile has a parabolic shape, and light is guided along the fiber due to refraction and self-focusing. The light path has a helicoidal shape with a period of Λ_0 (in the millimeter range). In a gradient-index fiber with a Gaussian-index profile, multimode dispersion is widely reduced.

Figure 1.28 shows another important quantity of optical fibers: the acceptance angle, θ_0. This angle is the maximum angle under which light can enter the fiber so that it still fulfils the condition of total reflexion at the core–cladding interface. Thus, $2\theta_0$ is the maximum aperture angle of the light cone that is allowed to enter and propagate through the fiber. From the acceptance angle, one defines the numerical aperture (NA) of the fiber, and it is a function of the refractive indices of surrounding medium (n_0), core (n_1) and cladding (n_2):

$$\text{NA} = n_0 \sin\theta_0 = \sqrt{n_1^2 - n_2^2} \tag{1.29}$$

As usual, the numerical aperture is a function of the light wavelength due to the dispersion of the fiber material. Table 1.10 shows a list of typical acceptance angles, numerical apertures, and transmission ranges for commercially available fibers [35]. Here, the numerical aperture can reach values of up to 0.86 which corresponds to an acceptance angle of 60°.

Several influences will cause light damping in an optical fiber. Material inhomogeneities smaller than the wavelength cause losses due to Rayleigh scattering. Absorption losses are caused by metal-ion impurities, and concentrations of 10^{-10}–10^{-9} yield a damping of about -1 dB/km. OH$^-$ ions with their wide absorption bands in the (near-) infrared lead to a strong damping and define transmission windows at wavelengths of 850 nm, 1300 nm and 1600 nm. Reflection losses occur due to irregularities between core and cladding. Apart from mode dispersion, there is, of course, also material dispersion. Its influence can be minimized by employing monochromatic light sources such as lasers or LEDs.

Although the principle of total internal reflection prohibits the existence of fields outside the fiber core, there is an exponentially decreasing electrical field, the so-called evanescent field, that extends to very small distances into the fiber cladding and which oscillates with the frequency of the incident light wave. The existence of the evanescent field can be explained by the tunnel effect of quantum mechanics. When the cladding is removed from two fibers and when the two exposed cores are brought into close contact, the evanescent field from the signal-carrying fiber is transformed into a propagating wave, and light energy is coupled into the other fiber. This is the basic principle of fiber couplers with which the distribution of light signals in fiber networks becomes possible.

4.7
Modulators

Many sensor applications work with continuous illumination. There are cases, however, in which it is necessary to modulate the light source. Although a modulation can be generated directly, for example, by modulating the driving current of an LED, this may either be impossible or undesired in certain applications.

The easiest way to realize an amplitude modulation is the employment of a so-called light chopper: a rotating, blackened disk carries regularly spaced holes. When rotating at a constant speed, a light beam will periodically be switched on and off. An additional light barrier that picks up the same frequency can then be used as a convenient reference input for a lock-in detection of the signal. The size of the disk and its maximum rotation frequency will, however, strongly limit the maximum possible modulation frequency.

A simple method to modulate the frequency of a laser beam makes use of the Doppler effect: Light emitted by a moving source will be detected with a higher frequency, that is, with smaller wavelength, when the source moves toward the detector, and with a longer wavelength when it moves away from it. This is not necessarily associated with a moving light source, but it may also be realized in the form of an oscillating mirror that reflects an incident light beam. Suppose that the mirror describes a harmonic oscillation along the optical axis with a displacement of $s(t) = s_0 \cdot \cos(2\pi f_{mod} t)$. From the resulting Doppler shift of the (angular) light frequency, $\Delta\omega = 2\pi \cdot \Delta\nu = 2k v_0$, ($k$ is wave number and v_0 the maximum oscillation speed), and with the basic relation for harmonic oscillations: $v_0 = 2\pi f_{mod} s_0$, it follows for the relation between the oscillation stroke, $2s_0$, and f_{mod} that

$$2s_0 \, f_{mod} = \frac{\lambda \cdot \Delta\nu}{2\pi} \tag{1.30}$$

This equation is valid for one beam reflexion on the oscillating mirror. If, however, the beam is reflected N times on the mirror, the stroke for a given

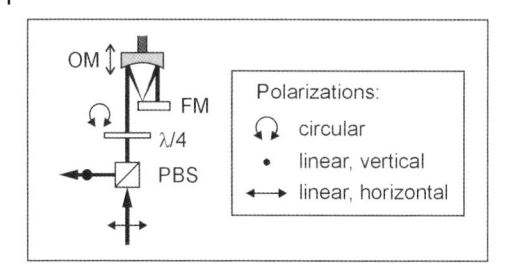

Figure 1.29 Oscillating-mirror modulator. OM: oscillating mirror, FM: fixed mirror, $\lambda/4$: quarter-wave plate, PBS: polarizing beamsplitter.

modulation frequency and a given Doppler shift (e.g., for a given frequency stroke in the spectrum) reduces to the $1/N$-th part of the value obtained from Eq. (1.30).

This purely mechanical technique has a bandwidth of less than 1 kHz, but this is already enough for sweeping a laser frequency across narrow atomic lines in high-resolution laser spectroscopy [36]. In its optical setup (Figure 1.29), the incident, horizontally polarized laser beam passes a polarizing beamsplitter, almost without reflective losses. A quarter-wave plate changes the state of polarization from linear to circular. The beam is then reflected by the oscillating, spherical mirror onto a fixed, plane mirror that is placed in the focal plane of the spherical mirror. Reflected back onto the spherical mirror and again onto the fixed mirror, the path of the returning beam is the same as for the incident beam. As the direction of propagation is inverted by the mirrors, the direction of the circular polarization is inverted as well. Thus, the quarter-wave plate changes the state of polarization back to linear, but now perpendicular to that of the incident beam. As a consequence, the beam is now reflected at right angles by the polarizing beamsplitter, and can now be used for experiments that require frequency-modulated laser light. The oscillating mirror is spherical, in order to prevent any beam displacements in the case of possible wobbling movements.

Electrooptical modulators (EOMs) are based on the Kerr effect or on the Pockels effect. These effects describe birefringence induced by electrical fields in isotropic media. The electrical field defines an optical axis that points into the direction of the electrical field vector. Thus, like in the case of natural birefringence, a difference, Δn, develops between the refractive indices of ordinary and extraordinary light, and it is a function of the applied electrical field, E [1]:

$$\Delta n = \lambda_0 \, K \, E^2 \tag{1.31}$$

Here, λ_0 denotes the vacuum wavelength and K is the Kerr constant. The induced birefringence leads to a phase difference between ordinary and extraordinary light. When leaving the medium, the two components interfere again and form an outgoing polarization that is rotated with respect to that of the incoming light. EOMs working on this principle consist of a Kerr cell that contains a polar liquid, for example, benzene, chloroform, or water. This cell is usually placed between two polarizers oriented under $\pm 45°$ with respect to the electrical field. When no voltage is applied, the light path is blocked, and with an applied voltage, the light path is open. These modulators can be operated at frequencies up to

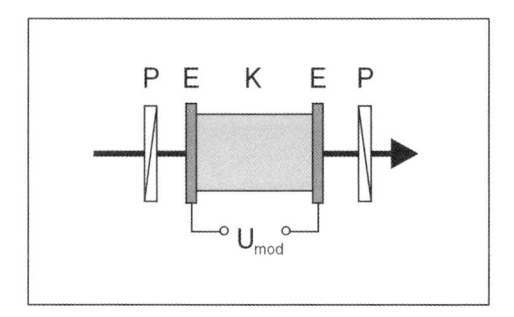

Figure 1.30 A Pockels cell. P: polarizer, E: transparent electrode, K: KDP crystal, U_{mod}: modulation voltage.

10^{10} Hz, and are therefore, for example, employed as ultrafast shutters in high-speed photography.

While the Kerr effect is an effect proportional to the square of the electrical field, there is also a linear electro-optical effect, the Pockels effect. It is found in crystals without centrosymmetry. With the refractive index for ordinary light, n_0, the electro-optical constant r_{63} (given in m/V) and the modulation voltage U_{mod}, the phase shift $\Delta\Phi$ between ordinary and extraordinary light is [1]:

$$\Delta\Phi = \frac{2\pi \, n_0^3 \, r_{63} \, U_{mod}}{\lambda_0} \tag{1.32}$$

One typical material that exhibits the Pockels effect is KH_2PO_4 (potassium dihydrogen phosphate, or in short: KDP). Its response time is less than 10 ns so that modulation frequencies of some 10 GHz are possible. In the longitudinal arrangement of a Pockels cell, the electric field is applied via transparent electrodes on the crystal faces (Figure 1.30). Pockels cells are used as fast shutters, for example, as so-called Q-switches in pulsed lasers [14].

Magneto-optical modulators work in analogy to EOMs, only that the plane of polarization is periodically rotated by a magnetic field. With a longitudinal magnetic field, this is a technical application for the well-known Faraday effect. With a transverse magnetic field in gases, the corresponding effect is the Voigt effect, in liquids, the Cotton–Mouton effect. A discussion of these effects can be found in [1].

An acousto-optical modulator (AOM) is a device in which light interacts with acoustic waves. Its design is based on a transparent material like glass or quartz with a piezo crystal as acoustic transducer. Driven by an external signal source, this transducer will send a propagating density wave into the material. This density wave corresponds to a propagating refractive index wave. The modulation frequencies can reach some hundreds of kHz, or even some GHz. In water, where the speed of sound is around 1 km/s, a frequency of 1 GHz will result in a wavelength of about 1 µm. This wavelength can be understood as the grating constant of an acoustically generated phase grating. Figure 1.31 shows a sketch of an AOM's working principle.

The phase grating acts as a diffraction grating with grating constant g. The output interference orders can be described in analogy to Eq. (1.23). With the angle

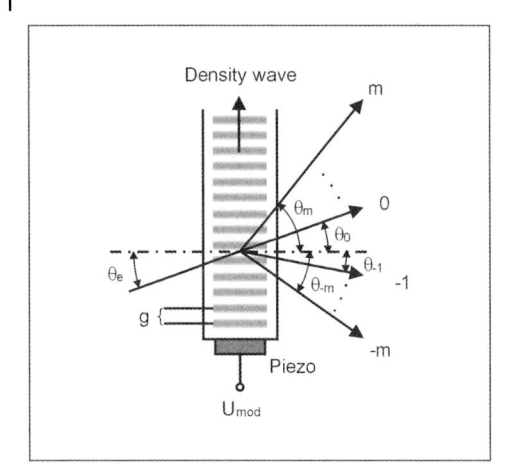

Figure 1.31 Acousto-optical modulator (AOM). U_{mod}: modulation voltage, g: grating constant. θ_e: angle of incidence. The indices m of the output rays under the diffraction angles θ_m indicate the respective interference orders.

of incidence, θ_e, and with the grating constant g, the diffraction angle θ_m of the mth diffraction order inside the medium is

$$\sin\theta_m = \sin\theta_e + \frac{m \cdot \lambda}{g} \quad m = 0, \pm 1, \pm 2, \pm 3, \ldots \tag{1.33}$$

The angles outside the medium can easily be obtained by multiplying θ_m with the index of refraction of the medium. Thus, the AOM can be used for the deflection of laser beams. As this deflection is caused by moving grating lines, the output beams are also frequency modulated due to the additional Doppler shift. In this application, the AOM is also denoted as a Bragg cell. Additionally, the intensity of the beam in each diffraction order depends on the intensity of the sound wave. Thus, an AOM can also be used for amplitude modulation.

References

1 Hecht, E. (1989) *Optik*, Addison-Wesley, Bonn.
2 SLI Sylvania (2002) Halogen Lamps, Product Catalogue.
3 Mayer-Kuckuk, T. (1985) *Atomphysik*, Teubner, Stuttgart.
4 Messiah, A. (1999) *Quantum Mechanics*, Dover, Mineola, NY, USA.
5 Osram GmbH (2004) Mercury Short Arc Lamp HBO® 50 W/3, Technical Information No. FO 4635.
6 Venture Lighting International (2006) North American HID Lighting Specification Guide, Product Catalogue.
7 Kittel, C. (1988) *Einführung in die Festkörperphysik*, Oldenbourg, München, Wien.
8 Nichia Corporation (2007) Specifications for Chip Type UV LED, Model NCSU034A(T), Product Datasheet.
9 Philips Lumileds (2007) Technical Datasheet DS51: Power Light Source LUXEON® K2, Product Datasheet.
10 EPIGAP Optoelektronik (2007) Jumbo LED ELJ-850-629, Product Datasheet.
11 Visser, P. (2008) *Optik–Photonik*, 3, pp. 27–29.
12 Shidlovski, V. (2004) Superluminescent Diodes. Short Overview of Device Operation Principles and Performance Parameters, Product Information SuperlumDiodes, Moscow, Russia.
13 Superlum (2009) Online Catalogue of SuperlumDiodes, Moscow, Russia, www.superlumdiodes.com (accessed 13.11.2009).
14 Kneubühl, F.K. and Sigrist, M.W. (1999) *Laser*, Teubner, Stuttgart.
15 Sanyo (2007) Blue-Violet Laser Diode DL-5146-101S, Tentative Product Datasheet.
16 Roithner Lasertechnik (2009) Online Catalogue of Roithner Lasertechnik, Vienna, Austria, www.roithner-laser.com (accessed 13.11.2009).
17 Laser 2000 (2005) Laserquellen:Helium-Neon-Laser, Product Datasheet.
18 Giessen, H., Hoos, F., and Teipel, J. (2005) *Themenheft Forschung (University of Stuttgart) – Photonics*, 2, 2–7.
19 Koheras (2009) SuperK(TM) White-Light Lasers, Product Documentation.
20 v. Ardenne, M., Musiol, G., and Reball, S. (1990) *Effekte der Physik und ihre Anwendungen*, Harri Deutsch, Thun, Frankfurt/Main.
21 Hakamata, T. (ed.) (2006) *Photomultiplier Tubes, Basics and Applications*, Hamamatsu Photonics K. K., Shizuoka-Ken, Japan.
22 Burle (2003) Channeltron: Electron Multiplier Handbook for Mass Spectrometry Applications, Product Documentation.
23 PerkinElmer (2003) TPS 534 Thermopile Detector, Product Datasheet.
24 Coates, C. (2007) OLE Product Guide 2007, 7–9.
25 Guttosch, R.J. (2005) *Investigation of Color Aliasing of High Spatial Frequencies and Edges for Bayer-Pattern Filters and Foveon X3® Direct Image Sensors*, Foveon Inc., Santa Clara, CA.
26 Hamamatsu Photonics (2007) Si PIN Photodiode S5106, S5107, S7509, S7510, Product Datasheet.

27 Hamamatsu Photonics (2005) 13 mm Dia. Side-On Compact Type Photomultiplier Series, Product Datasheet.
28 Laser Components (2007) Silicon Avalanche Photodiode SAR1500x/3000x, Product Datasheet.
29 Naumann, H. and Schröder, G. (1992) *Bauelemente der Optik*, Carl Hanser Verlag, München, Wien.
30 Schott Advanced Optics (2007) Optical Glass – Datasheets. Product Information.
31 Döbler, M. (2008) Polycarbonat für optische Anwendungen, Product Information, Bayer MaterialScience, Business Unit Polycarbonates.
32 Palmer, C. and Loewen, E. (2005) *Diffraction Grating Handbook*, Newport Corporation, Rochester, NY.
33 Kaiser, N. (2006) *Photonik*, **6/2006**, 50–52.
34 Bass, M. (1995) *Handbook of Optics*, McGraw-Hill, New York.
35 Helmut Hund (2009) Fiber Optics – Applications and Options, Product Information.
36 Bernhardt, J., Haus, J., Hermann, G., Lasnitschka, G., Mahr, G., and Scharmann, A. (1998) Laserspektrometrischer Nachweis von Strontiumnukliden, in *Zivilschutzforschung, Neue Folge Band 33*, Schriftenreihe der Schutzkommission beim Bundesminister des Innern, Bundesamt für Zivilschutz, Bonn, pp. 47–51.

Part Two
Optical Sensors and Their Applications

This section is structured according to the properties that optical sensors measure. There will be a certain emphasis on mechanical properties, but the examples will give the reader an impression about how versatile sensor principles can become.

Optical Sensors: Basics and Applications. Jörg Haus
© 2010 WILEY-VCH Verlag GmbH & Co. KGaA, Weinheim
ISBN: 978-3-527-40860-3

5
Eyes: The Examples of Nature

While technical optical sensors emerged with the advent of the first optical detectors, that is, elements that convert optical to electrical signals, the first optical sensors ever, on the other hand, have been developed by nature itself. They are the product of millions of years of evolution: the eyes of biological organisms. These can be grouped into three different types:

- primitive eye pits without any lens-like elements,
- compound eyes composed of thousands of single eyes,
- eyes that project an image with one central lens.

The latter two types will be discussed in more detail.

5.1
The Compound Eyes of Insects

A compound eye is a hexagonal arrangement of thousands of single eyes, so-called ommatidia. Each ommatidium delivers one picture element to the brain, and their number ranges from just a few hundred per eye to more than 20 000 per eye for dragonflies.

The ommatidia are closely packed, and their cross sections decrease from the outer surface inward (Figure 2.1). Light enters through the cornea lens and is further focused by the pseudocone. The inner section contains six or seven photosensitive cells, the "R" cells. On the central axis of the ommatidium, these cells form a light guide, the so-called rhabdom. Each photonsensitive cell is connected to the nervous system by one axon. To prevent crosstalk between the ommatidia, their perimeters are lined with pigment cells. This has the drawback that only a limited amount of light can be received, but in low-light conditions, the pigment cells of some insect species can be retracted. This increases the light sensitivity by increasing the effective "pixel size," but reduces the resolution of the compound eye.

The photoreceptors of insects are not only sensitive in the visible spectrum, but also in the ultraviolet. The receptors are mono- or tetrachromatic with peaks at 340 nm (UV), 450 nm (blue), 500 nm (green), and 620 nm (red) [1]. Not all insect species, though, have blue and red receptors.

Optical Sensors: Basics and Applications. Jörg Haus
© 2010 WILEY-VCH Verlag GmbH & Co. KGaA, Weinheim
ISBN: 978-3-527-40860-3

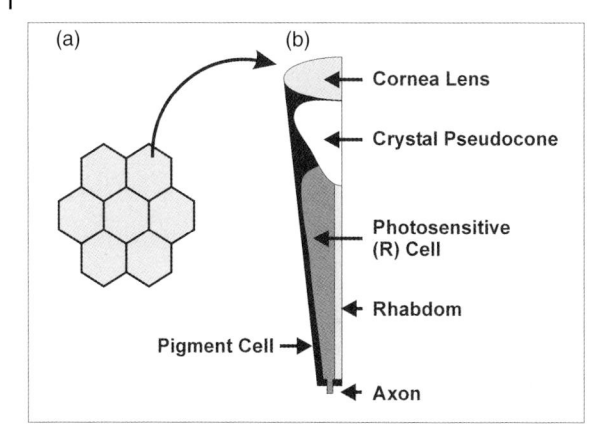

Figure 2.1 Hexagonal structure of an insect eye (a), simplified illustration of a single ommatidium (b).

As compound eyes are designed as discrete receptor arrays, they are very sensitive to changes in the image, that is, they are motion sensitive. This design is very effectively mimicked in all sensors that measure speed based on optical cross-correlation (see Section 6.2.3).

5.2
Nature's Example: The Human Eye

The human eye is a spherical organ with a diameter of a little less than 25 mm. Its optical setup closely resembles that of a pinhole camera: light enters through the cornea and the entrance pupil, and the image is formed by a single lens on the retina, the neural layer that contains the photoreceptors. There are rods for the perception of low-light contrasts (scotopic vision) and cones for color (photopic) vision (Figure 2.2). The optic nerve then transfers the neural signals to the visual cortex of the brain. The space between lens and retina is filled with a transparent, colorless mass with high viscosity, the vitreous humor. It mainly consists of water (99%), sugar, salt, and collagen fibers. It contains the hyaloid canal, a relic of the hyaloid artery that connects lens and optic disk in the human fetus.

The eye is encapsuled by the sclera, which is white and opaque except for its front section, the transparent cornea. Its index of refraction is almost 1.4 so that refraction occurs mostly at the air–cornea interface. The entrance pupil is formed by the iris, and its color determines the characteristic appearance of an individual's eyes. The diameter of the iris varies with the incident light intensity, and it also helps to reduce chromatic aberrations of the single-lens system. The lens itself consists of more than 20 000 layers across which the index of refraction changes from 1.406 to 1.386 from the outer shell to the core [3]. The lens can continuously focus onto objects in different distances by deforming as a whole by means of a ring muscle.

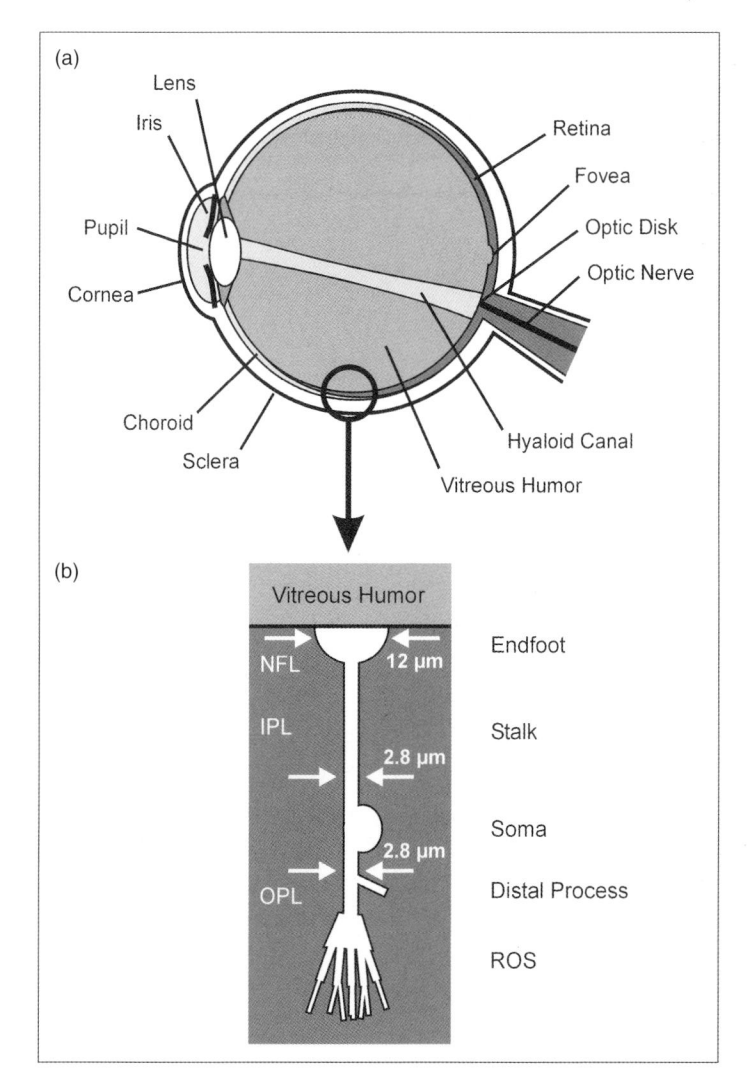

Figure 2.2 The human eye. (a) Cutaway view, (b) detail of the retina with Müller cell (white) (adapted from [2]). Arrows: Müller cell diameter at three different depths, abbreviations indicate singular layers of the retina. NFL: nerve fiber layer, IPL: inner plexiform layer, OPL: outer plexiform layer. ROS: receptor outer segments.

Humans are trichromats in the sense that about 120 million cones of the retina have absorption maxima in three different parts of the visible spectrum: "red" cones have their absorption maximum at around 565 nm, "green" cones at around 535 nm, and "blue" cones at around 420 nm. As a result, the human eye has an overall daylight sensitivity curve, $V(\lambda)$, with a maximum at 555 nm (Figure 2.3). On the other hand, about 6 million rods determine the visual impression under

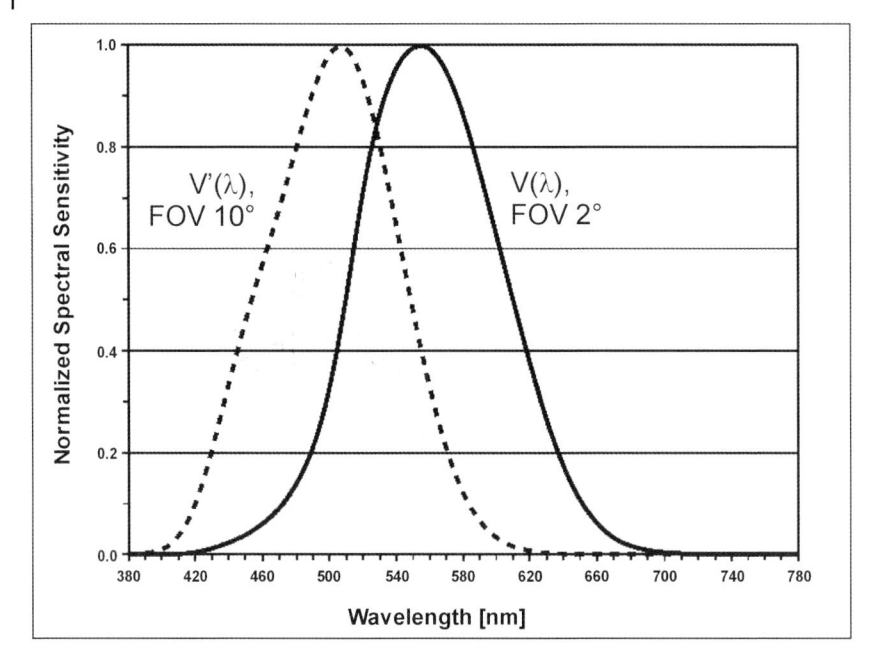

Figure 2.3 The Purkinje effect: Shift of the human eye's spectral sensitivity curve from photopic vision, $V(\lambda)$, to scotopic vision, $V'(\lambda)$. FOV: field of vision.

low-light conditions. Their spectral sensitivity is highest at around 498 nm so that the weighted overall sensitivity, $V'(\lambda)$, now peaks at 507 nm ([4], Figure 2.3). This dark-adaptation process is also denoted as the Purkinje shift.

The distribution of rods and cones across the retina is not uniform: While the rods have their highest concentration on the periphery, the concentration of the cones is highest at the center, and vice versa. The central area where the cone concentration is highest is the fovea; the region of sharpest vision. This explains why objects under twilight conditions are best observed indirectly, but then, with high sensitivity.

The optic disk is the spot where the optic nerve terminates. Here, the neural fibers enter the inside of the eyeball and connect to the retina. As a consequence, there are no photoreceptors here and the eye is completely blind at this spot. The existence of the optic disk points to another interesting property of the human (and also of the vertebrate) retina. Its design is "inverted," which means that the photoreceptors are arranged on the backside of the retina's neural tissue. Although this design can be explained from the evolutionary development of the human eye, it does not seem to be very reasonable from the point of view of an optical designer: As the light has to pass the whole of the retinal tissue before it can be detected, the image would actually be blurred by scattering and diffraction of light. However, it was found that the retina contains so-called Müller cells, which seem to act like light guides that bridge the retinal tissue [2]. Thus, the retina acts like a fiber-optic image plate, and the neural tissue in front of the photoreceptors does not affect the image.

All vertebrates and humans have one pair of eyes that are horizontally arranged in their skulls. Both eyes thus capture slightly different, overlapping images of a scene. Also, objects in different distances are observed under different angles. This binocular vision enables the brain to perceive depth distributions in a scene and also to estimate distances to objects.

6
Optical Sensor Concepts

The examples in this section are arranged according to the properties that the sensors measure. Its purpose is to give an impression of the versatility of optical sensors and of how physical properties are translated into optical working principles.

6.1
Switches

A switch is an element with only two output states: ON and OFF or, digitally spoken, 0 and 1. Optical switches can easily be realized with an emitter–detector pair. Every object passing this optical channel will interrupt the light beam and trigger a signal pulse in the detector. A light barrier is a comparably simple example for an optical switch, but a somewhat more sophisticated embodiment is the rain sensor used in modern-day automobiles.

6.1.1
Light Barriers

Light barriers are probably the most widely employed optical sensors in all fields of technology. Their task is, for example, to start or stop measurements, to set start/stop points in process control, or to initiate safety protocols.

The construction of light barriers is comparatively simple – a light source, either a diode laser or an LED, emits light that is again detected by a photodetector. Depending on the respective application, light barriers can be transmissive or reflective (Figure 2.4). With transmission-type light barriers, an object that interrupts the light path gives rise to an impulse in the detection circuit. For reflexion-type light barriers, two designs are possible: in the first, any object in the light path will scatter or reflect light back into the detector, in the second, the object will interrupt the light path. Reflexion-type light barriers have advantages over transmission-type ones for large path lengths and in dynamic processes, and light source and detector can be included in the same housing.

Optical Sensors: Basics and Applications. Jörg Haus
© 2010 WILEY-VCH Verlag GmbH & Co. KGaA, Weinheim
ISBN: 978-3-527-40860-3

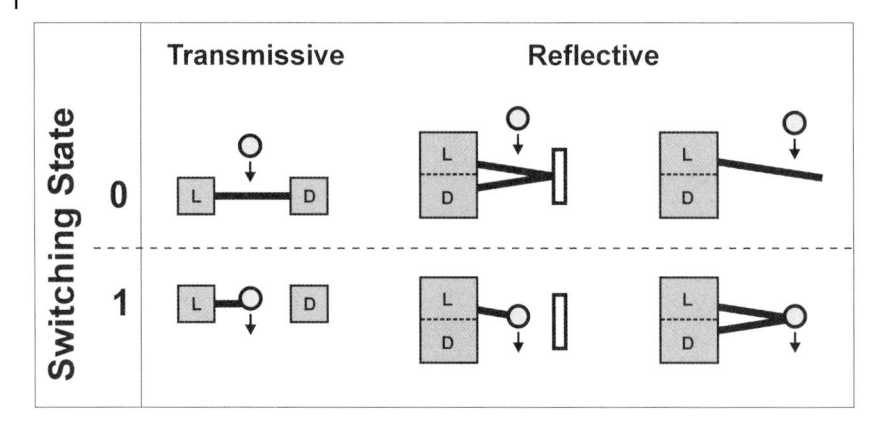

Figure 2.4 Light barrier designs. L: light source, D: detector. Left: transmission-type, right: reflexion-type.

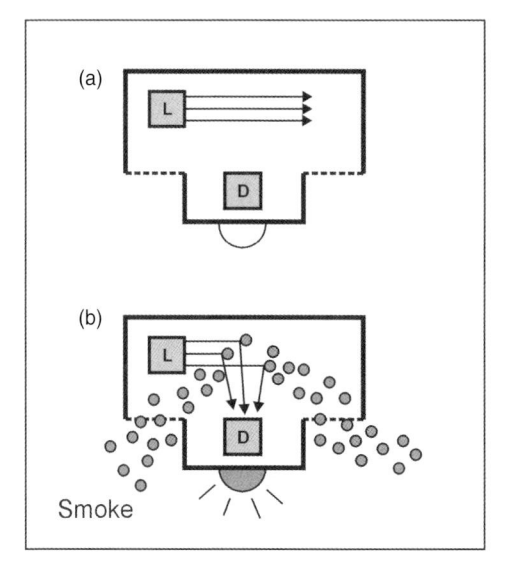

Figure 2.5 Smoke detector. L: light source, D: detector. (a) with clean air, (b) with smoke.

Both diode lasers and LEDs may be employed as light sources. With appropriate optical designs, light barriers can bridge light paths of several hundred meters in commercially available systems.

One special example of a reflexion-type light barrier is a smoke detector. It makes use of the scattering of light by smoke particles in a 90° scattering geometry (Figure 2.5). When the air in the detector chamber is clean, no light from the diode laser will reach the detector. In the presence of smoke, however, the particles in the chamber will scatter light into the direction of the photodiode, and the alarm will go off. When not only the presence of a signal is evaluated, but also its

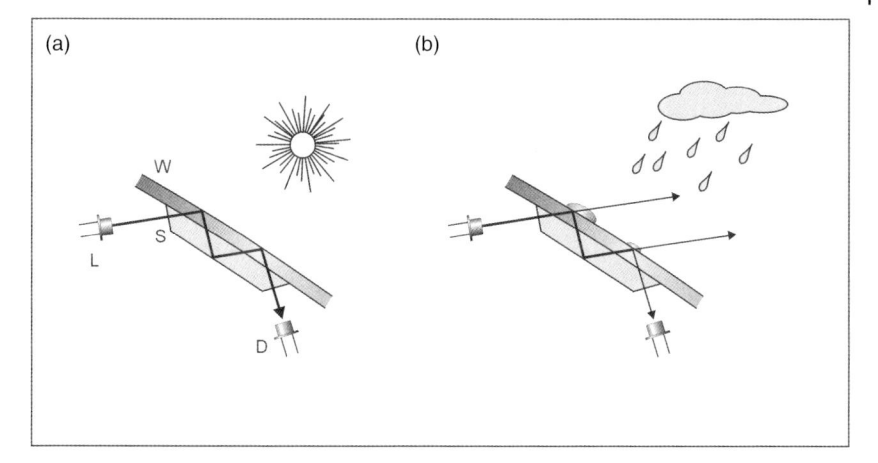

Figure 2.6 Rain sensor. L: light source, S: sensor window, W: windshield, D: photodetector. (a) Dry weather, (b) rainy weather.

magnitude, this arrangement can also be used as a sensor for particle mass concentrations, see Section 6.5.5.

6.1.2
Rain Sensor

A modern accessory for motorcars is the rain sensor for the automatic activation of the windscreen wiper when it rains. Its working principle is based on the disturbed total internal reflection of light on the windscreen–air interface (Figure 2.6).

An LED emits light into the material of the windscreen. Whenever light travels from a medium with refractive index n_i to a medium with smaller refractive index n_R, the angle of refraction is larger than the angle of incidence according to Snellius' law. For the critical angle of incidence, θ_C, the angle of refraction, θ_R, reaches 90°. For angles larger than θ_C, light is completely reflected from the interface back into the windshield. For the critical angle, it is

$$\sin\theta_C = \frac{n_R}{n_i}\sin\theta_R = \frac{n_R}{n_i} \quad (n_R < n_i) \tag{2.1}$$

Under dry weather conditions, the windscreen thus acts as a waveguide from which no light can exit, and it directs the light onto the photodetector.

When raindrops, snow, or dew hit the windscreen, n_R changes to a value close to n_i. As a consequence, the condition of total internal reflection is disturbed. Light can leave the windshield, and the intensity received by the photodetector decreases. The resulting signal is then used for the automated control of the windscreen wipers.

6.2
Spatial Dimensions

There are numerous ways to measure spatial dimensions with nonoptical techniques. Optical techniques, however, generally have the advantage of working in a noncontact manner so that measurement errors like those introduced by slip are effectively prevented.

This section discusses optical sensors for measuring distances between the sensor and some distant position, displacements of two objects relative to each other, velocities relative to a moving surface, and angular velocities.

6.2.1
Distance

The distance between a sensor and an object is an important quantity for a large number of applications, when aiming at a target, or when adjusting the sharpness of an optical image; in short, whenever information about the relative positions of objects is required.

The simplest way to determine the distance to an object by means of light is to exploit the definition of the meter: 1 meter is the distance that light travels in vacuum during a time interval of $(1/299\,792\,458)\,\text{s} \approx 3.34\,\text{ns}$ [5]. Thus, it takes light only about 3.34 ps to bridge a distance of 1 mm! This means that time-of-flight measurements require very sophisticated sensor setups with extremely fast electronic circuits to achieve high spatial resolutions. Imagine a laser pulse that travels to a reflector in some distance, L, and then the same distance back to a detector. The time for one round-trip of the pulse, Δt, is (c is the speed of light):

$$L = \frac{c \cdot \Delta t}{2} \tag{2.2}$$

One workaround for the necessity to measure extremely short time intervals is to modulate the laser beam with some frequency, f_{mod}. The distance that the light has traveled will then result in a phase shift, θ, of the modulation signal that the detector receives. This phase shift is proportional to the distance. With $T = 1/f_{mod}$, one gets

$$\theta = \frac{\Delta t}{T} \cdot 2\pi \tag{2.3}$$

The distance, L, is then calculated according to Eq. (2.2):

$$L = \frac{c \cdot \Delta t}{2} = \frac{cT}{4\pi}(\theta + n \cdot 2\pi) \tag{2.4}$$

Please note that an additional term $n \cdot 2\pi$ is added because of the periodicity of the modulation. The ambiguity of the measurement result must be corrected with additional components. Here, the necessary specifications of the electronic

Table 2.1 Technical specifications of a commercial LIDAR sensor [6, 7].

Property	Value
Measurement range	200 m
Regulation range	150 m
Viewing angle	16° (horizontal) × 3° (vertical)
Measurement channels	16
Resolution	0.1 m
Measurement uncertainty	1%
Operating temperature	−40 °C to +85 °C

circuit are determined by the modulation frequency, and not by the laser frequency.

Time-of-flight laser sensors are also called light detection and ranging (LIDAR) sensors. Commercial systems exist for use in driver assistance systems, for example, as distance sensors for adaptive cruise control (ACC) systems. One of these utilizes infrared light, has a viewing angle of up to 16° (horizontal) × 3° (vertical) and a measurement range of 200 m. A maximum of 16 measurement channels allows to determine an object's lateral position in the sensor's field of view. It reaches a resolution of 0.1 m and a measurement uncertainty of 1%. Additionally, all evaluation and ACC algorithms, for which it also requires input from the wheel sensors and steering angle, transverse acceleration, and yaw rate sensors are implemented in the LIDAR sensor [6, 7]. Table 2.1 shows an overview of its specifications.

One very sophisticated and highly integrated embodiment of this measurement technique is the photon-mixing device (PMD). As an active sensing concept, it requires that a scene be illuminated with modulated light. Based on standard CCD or CMOS architectures, every pixel of this device not only receives the scene as a 2D image, but its design also enables the PMD to extract time-of-flight or distance information by an integrated mixing of the received signal with the modulation input. This makes the PMD a true 3D sensor. The technology reaches frame rates up to 100 fps and is in the focus of research for various kinds of driver assistance systems in the automotive industry [8]. One example for a possible application is the "smart" airbag: The sensor detects if a seat is occupied, and if so, if the person on the seat is sitting upright. Thus, the airbag will be fired regularly, with some delay, or even not at all. The PMD technology becomes even more interesting as there are more and more LEDs installed in modern-day vehicles. They serve as daylight-running lights, indicators, tail lights, and even as headlights. The PMD concept may thus be integrated seamlessly into this context because the necessary modulation frequencies of the LEDs are too high to be perceived by the human eye, and so they cannot confuse the driver. Reference [9] describes the prototype of an automotive front view camera that may be used for, for example, distance

Table 2.2 Technical specifications of a PMD sensor prototype [9].

Property	Value
Measurement range	10 m (Light source power: 1 W)
Viewing angle	18° (horizontal) × 52° (vertical)
Measurement rate	60 Hz
Resolution	±0.1 m
Operating temperature	−10 °C to +85 °C

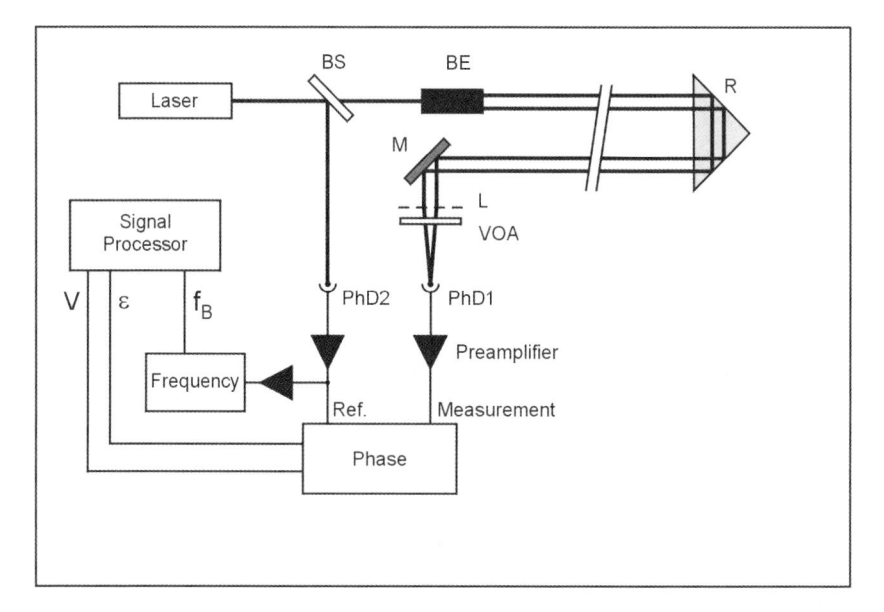

Figure 2.7 Laser interferometer. BS: beamsplitter, BE: beam expander, R: reflector, M: mirror, L: lens, VOA: variable optical attenuator, PhD: photodiode, V: signal intensity, ε: phase difference, f_B: beat frequency.

regulation in stop-and-go traffic situations or pedestrian safety applications. Table 2.2 shows its specifications.

As the distance information is contained in the phase shift between emitted and received light, this also offers the possibility to utilize interferometric techniques. Figure 2.7 shows an example (adapted from [10]): The light source is a wavelength-stabilized laser that oscillates on two longitudinal modes. These orthogonally polarized modes are mixed at the output, for example, with a polarizer. The laser light is thus modulated with a beat frequency of usually several hundreds of megahertz, corresponding to the mode spacing of the laser.

One half of the laser output, the measuring beam, is directed to a distant reflector and back to a photodiode. The other half, the reference beam, is directed to a second photodiode. The device measures the phase difference, ε, between the measurement and reference signals and the beat frequency of the laser modes, f_B. Then, the distance D between the laser setup and the reflector is calculated according to [10]

$$D = \frac{c(N+\varepsilon)}{2n_L f_B} + D_C \qquad (2.5)$$

with n_L is refractive index of air, N the unknown (large) integer, D_C the constant of the apparatus. While f_B and ε are measured directly, N has to be determined from two independent measurements with different beat frequencies. D_C contains the optical path length between the optical components and phase delays in the electronic circuits and can also be determined from measurements.

Laser interferometers are, for example, employed for the highly accurate calibration of coordinate measuring machines (CMMs). While the laser setup is realized in one single housing, the reflector is used as a remote target mounted to the CMM's moving frame. When it travels to different positions, the interferometer readings and those of the CMM's internal scales are compared in order to calibrate the internal scales. A typical interferometer [11] reaches linear resolutions of up to 0.25 nm and can measure distances of up to 80 m (Table 2.3). Special optical setups also allow the measurement of pitch and yaw angles during the CMM's motion along its axes and therefore allow the correction of all possible geometric errors of the CMM.

Moderate distances may be measured by recording the parallactic shift of a light spot on a surface. Here, the light source is a diode laser that emits perpendicularly to the surface, and the backscattered light is captured under some angle by the sensor's optics. Figure 2.8 shows a schematic setup of a triangulation sensor.

The detecting element is a PSD. When the distance between sensor and surface changes, the lateral position, Δx, of the light spot on the PSD changes. Thus, the PSD readout is a measure of the change in distance, $d' - d$. The longer the triangulation baseline, s (Figure 2.8), the higher the sensitivity of the sensor. In reality, however, this baseline may be limited due to dimensional constraints of the sensor housing.

Table 2.3 Technical specifications of a laser-interferometric distance sensor [11].

Property	Value
Measurement range	Up to 80 m (with long-range option)
Resolution (standard/extended)	2.5 nm/0.25 nm (with special optics)
Maximum axis velocity	Up to 0.25 m/s
Output power	$\geq 180\,\mu W$ (632.8 nm)
Operating temperature	0 °C to +40 °C

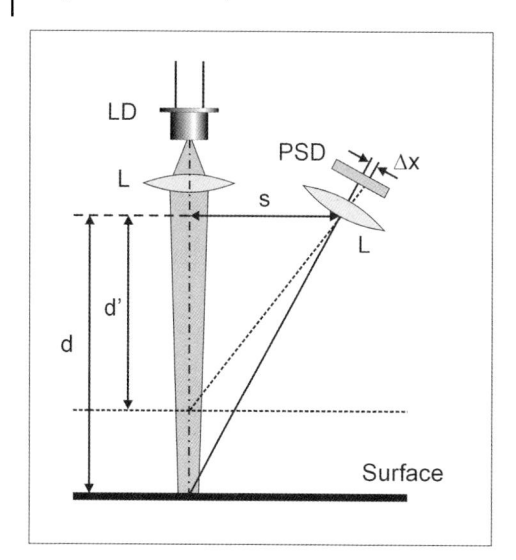

Figure 2.8 Laser triangulation sensor. LD: laser diode, L: lens, PSD: position-sensitive device.

Table 2.4 Technical specifications of a triangulation sensor [12].

Property	Value
Measurement range	100 m
Linearity	180 µm
Resolution	50 µm
Measurement rate	1 kHz
Operating temperature	0 °C to +55 °C

In principle, an initial measurement is necessary that relates the PSD's zero reading to an absolute distance to the surface. Then, triangulation sensors measure distance changes with linearity errors of around 1% and with reproducibilities of around 0.5% (see Table 2.4 for an example [12]). To operate properly, they require scattering surfaces, and specular reflections lead to significant intensity losses in the detection channel. Also, in dynamic measurements, the sensor will record the surface profile along its line of travel. As the lateral resolution is determined by the size of the laser spot, the sensor output may contain undesired, high-frequency components generated by the surface structure itself. This may require dynamic filtering of the sensor output.

A distance sensor that probably everyone is familiar with is the autofocus sensor of modern-day digital cameras. As contrast and sharpness of an image are linked together directly via the higher spatial frequency components of an image, optimization of the image contrast is a very direct method to realize an autofocus sensor. This requires no additional sensor besides the CCD or CMOS image sensor of the camera itself. More expensive solutions, for example, those in single-lens reflex cameras (SLRs), analyze two partial images of a scene, obtained by splitting the entrance pupil of the system with a prism pair. In focus, the partial images are identical, out of focus, they are different, but contain identical structures. Captured by line cameras, the cross-correlation between the images, i_1 and i_2, is (x representing the spatial coordinate along the line camera):

$$C(\Delta x) = \int_A i_1(x) \cdot i_2(x + \Delta x) dx \tag{2.6}$$

In principle, the shift between the images, Δx, is a measure for the focusing error. Focusing is the process of minimizing this error by maximizing the cross-correlation in Eq. (2.6), that is, by adjusting the optical system until the integral, calculated numerically and with discrete values, reaches its maximum. For a review on the history of the development of SLR autofocus systems, the interested reader is referred to Ref. [13].

6.2.2
Displacement

In contrast to distance, displacement is the property to be measured when two objects move in parallel planes, and relative to one another. This terminology may not be unambiguous in every case, but particularly when dynamic measurements are concerned, the measurement of displacement is the basis for the measurement of speed relative to some reference surface. Particularly in the dynamic measurements in vehicle testing, this is usually referred to as "speed over ground" (SOG).

High-precision displacement measurements rely on the relative motion of scales with periodical graduations. One typical application for these systems is a CMM (see above), a highly precise and accurate instrument that measures the dimensions of three-dimensional objects [14]. CMMs carry a probe head in a frame that allows its three-dimensional travel in a cartesian coordinate system. When the probe head gets in contact with the workpiece under investigation, the corresponding point in space is recorded. When then the probe head moves to a different point, its displacement is stored in the CMMs controller as one particular dimension of the workpiece.

Commonly, the displacement sensors are so-called incremental encoders [15]. The scales are reflection-type gratings with graduations of only a few 10 µm. They are either etched in sheet metal or glass rods, or even stamped on steel stripes. The sensor itself, the readhead, is mounted on a CMM component that moves parallel to the corresponding scale. It contains illumination LEDs, a reference grating, and a photodetector (Figure 2.9).

The condenser lens both acts as a collimator to illuminate the fixed scale and as a field lens to image the reference grating onto the detector. As the light passes the reference grating twice, this setup is also called a three-grating incremental

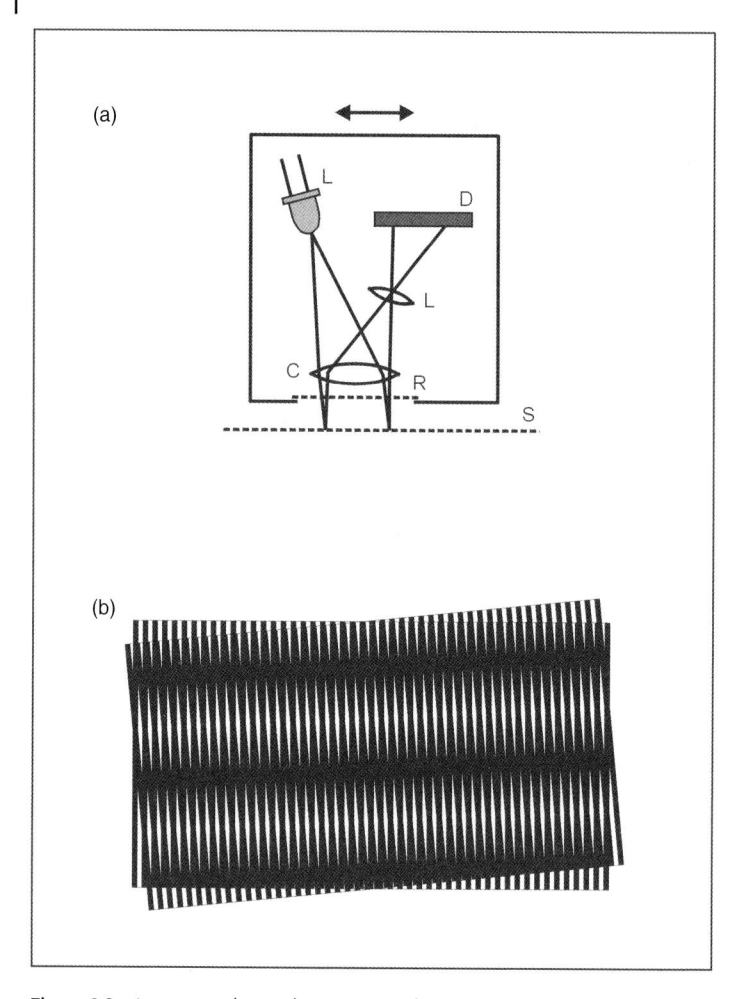

Figure 2.9 Incremental encoder. (a) Optical setup (schematic). L: light source, C: condenser lens, R: reference grating, S: scale, L: lens, D: detector. (b) Moiré pattern of two tilted gratings of identical graduations.

encoder. When scale and reference grating have slightly different grating constants, or if the reference grating is slightly tilted, a Moiré pattern forms in the detector plane (Figure 2.9). As the fundamental spatial frequency (i.e., the periodicity) of the Moiré pattern is simply the beat frequency of the spatial frequencies of the two patterns, it is much smaller than the spatial frequency of the scale itself. Thus, the Moiré pattern, which moves almost perpendicular to the movement of the readhead, is easy to detect and the number of dark lines that pass the detector is proportional to the path that the readhead has traveled with respect to the scale; that is, it is proportional to its displacement.

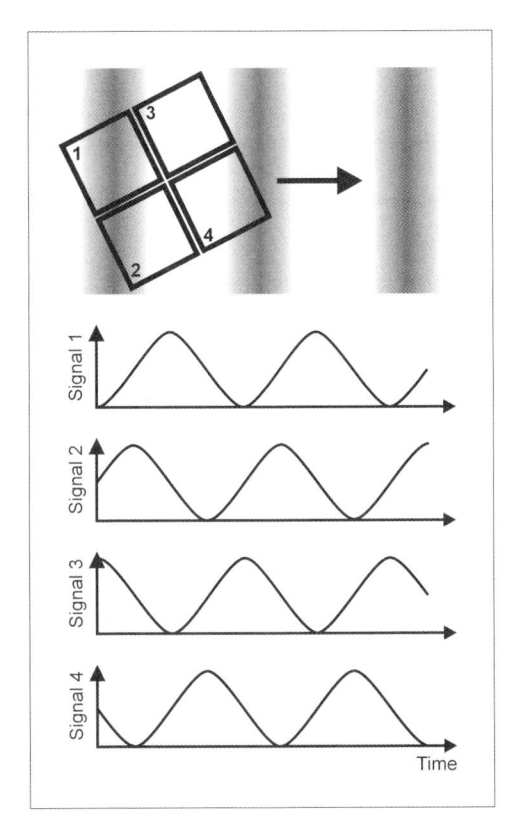

Figure 2.10 Incremental encoder: quadrature signals from a four-quadrant detector, tilted against the reference scale.

Interpolation techniques electronically increase the resolution of the encoder so that it reaches some 0.1 µm. This is realized by deriving several phase-shifted signals from the original one in an electronic network. Amplitude, offset, and phase errors of these single signals give rise to interpolation errors whose periods equal the pitch of the scale and can reach amplitudes comparable with the encoder resolution. This may require additional compensation algorithms.

Incremental encoders can also determine the direction of the displacement. This requires a quadrature signal, that is, two signals with a phase shift of 90°. One possible solution is the employment of two reference gratings whose graduations have a relative shift of one-fourth of a grating period. Another solution is based on a four-quadrant detector, a photodetector with four separate photodiodes, arranged in a square pattern (Figure 2.10, [15]).

As the detector is tilted with respect to the Moiré pattern, bright and dark lines cross the four single detectors with a certain time delay when the readhead moves. This delay depends on the tilt angle. For a quadrature signal, this delay has to be adjusted to yield a phase shift of 90° between the four signals, which are usually denoted by A (Signal 1), \A ("inverted A," Signal 3), B (Signal 2), and \B ("inverted B," Signal 4, Figure 2.10). As these signals are basically intensities, they contain offsets that can be compensated by subtracting A from \A and B from \B. When L is the distance between the centers of the single photodiodes, the relative phase shifts of 90° demand that the coordinates of the four photodiode centers in the direction of movement are spaced at equal distances of one-fourth of the distance of the Moiré fringes. Then, the tilt angle, γ, can be calculated to

$$\gamma = \arctan\frac{L/2}{L} = \arctan(0.5) \approx 26.6° \qquad (2.7)$$

When finer graduations are required, incremental encoders utilize diffraction effects for the generation of a periodical interference pattern. For more and in-depth information about the optical principles of incremental encoders, the interested reader is referred to Ref. [15]. A typical commercial linear encoder has a scale made of Zerodur glass ceramic with a grating constant of 0.512 μm. The readhead yields a signal period of 0.128 μm due to a fourfold interpolation based on the quadrature signals. A further 32-fold interpolation finally yields a signal period of 0.004 μm = 4 nm [16] (Table 2.5).

It is interesting to note that for every CMM, a calibration procedure is necessary to correct geometrical errors that build up due to linear, roll, pitch, and yaw deviations when the CMM moves. Also, the angles between the three machine axes may not exactly be 90°. Every CMM will therefore have to be corrected with a highly accurate length measurement system, usually based on a laser interferometer (see previous section). It should also be mentioned that besides incremental encoders, there are also rotary encoders that measure angular displacements. They work on a similar principle; the scales, however, are not linear, but arranged on a rotating disk.

When the employment of a reference scale is impossible or undesirable, the possibly easiest way to determine the displacement of an object relative to some reference is to take two images of this reference before and after the object has moved. When the scale factor of the imaging optics is known, the displacement can be calculated. The challenge here lies in the automated identification of the corresponding features in the two images. Probably everyone who works with a

Table 2.5 Technical specifications of an incremental encoder [16].

Property	Value
Scale	Zerodur glass ceramic
Grating period	0.512 μm
Signal period	0.128 μm/0.004 μm (4-/32-fold interpolation)
Measurement uncertainty	±0.5 μm
Maximum velocity	7.6 m/min (4-fold interpolation)
Operating temperature	0 °C to +40 °C

Table 2.6 Technical specifications of a mouse sensor [18].

Property	Value
Resolution (default)	About 20 counts/mm (500 counts/inch)
Maximum velocity	About 400 mm/s (16 inches/s)
Path error	0.5% at a path length of 63 mm (2.5 inches)
Operating temperature	−15 °C to +55 °C

personal computer knows an optical sensor that works on this principle – it is the heart of each optical mouse. Its complete optoelectronic setup is contained in one IC: an LED or laser diode with illumination optics, a fast camera that records more than 1.500 images per second, and an integrated image processing system with a USB interface [17]. This high integration makes it very easy to build a complete mouse around one single chip [18]. Although these mice are quite cheap today, the measurement principle also has its limitations – it is mandatory that the surface on which the mouse moves has a random structure that the camera can capture with sufficient contrast. This means that the mice cannot be used on completely smooth surfaces, for example, on glass plates. Mouse sensors with laser diodes (VCSELs) instead of LEDs help to improve the detection on surfaces with weak contrasts because their high intensities yield bright highlights that can easily be detected. Periodically structured surfaces will also be problematic for the sensor because subsequently recorded images will not unambiguously give sufficient information about the true displacement. The specifications of one typical mouse sensor can be found in Table 2.6.

Also, the maximum speed with which the sensor can be moved is given by the maximum frame rate and the size of the camera chip: if the speed is too large, two subsequent frames will not contain corresponding surface features any more. Reference [18] gives the maximum speed as about 40 mm/s (16 inches/s) and the maximum acceleration as 2g. In the same reference, the average path error along a path length of 2.5 inches is given as 0.5 %. However, this property is not very decisive for an optical mouse because the user will always close the "positioning loop" by means of visual information from the computer screen.

6.2.3
Velocity

Velocity is the rate by which the displacement, L, of an object changes over time. The well-known physical definition is

$$v = \frac{dL}{dt} \tag{2.8}$$

With an additional clock on board, a displacement sensor becomes a velocity sensor, and all techniques described in the last section will also be applicable here.

6 Optical Sensor Concepts

As we have seen, the mouse sensor has a limited velocity range. Higher speeds (or larger displacements over time) can be measured with a similar working principle that is based on an arrangement of two sensors at some distance L. These sensors do not yield a two-dimensional image of the surface, but transform the surface structure into a one-dimensional signal, for example, by recording the brightness of backscattered light or by measuring the surface profile with a triangulation sensor. The spatial distance, L, will then lead to a time delay, τ, in the two signals, i_1 and i_2. As L is given by the geometrical arrangement of the sensors, the measurement of velocity is reduced to a measurement of the time delay. This is usually done very similar to Eq. (2.6) by calculating the time-domain cross-correlation of the two signals and by maximizing the result (Eq. 2.9).

$$C(\tau) = \int_A i_1(t) \cdot i_2(t+\tau) dt \qquad (2.9)$$

This measurement principle can be employed in measuring the velocities of road vehicles and of industrial strip materials like paper webs, rubber, textiles or sheet metal (Figure 2.11). Depending on the parameters of the sensor setup and the surface, it is possible to reach measurement errors of less than 1% [19].

One disadvantage of these systems is that the time that a measurement value takes to be acquired equals at least the time interval that a certain surface spot takes to travel from one sensor to the other. A real-time measurement of low velocities, as may be a demand in dynamic measurements of railway vehicles, may thus be difficult with this system. The solution would be a reduction of the distance between the two sensors. As they point vertically down to the surface, however, there is a geometrical limit to the sensor distance. On the other hand, adjusting the sensors to view closely spaced fields under some oblique angle will cause difficulties, for example, in case of distance changes between sensor and surface.

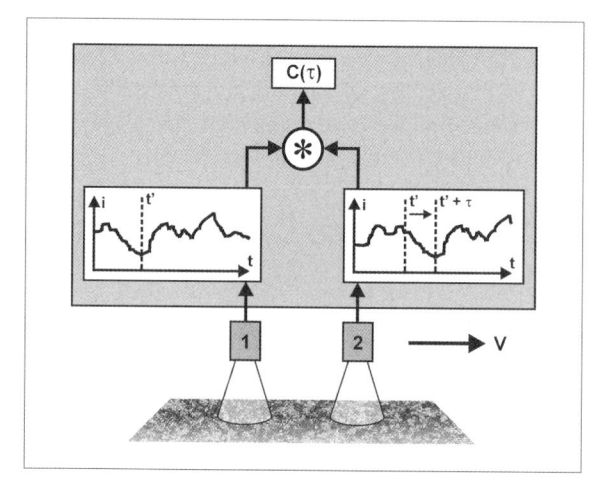

Figure 2.11 Time-domain correlation sensor. τ: time delay between the two sensor signals.

A very old technique for determining the velocity of an object is based on optical gratings (see [20] and references cited therein). An image of the object is projected onto an amplitude grating, and when the grating constant corresponds to the size of structures in the image, each of these structures will yield a periodical intensity signal on the detector behind the grating. This is an effect that can be observed quite easily when a light source moves behind a picket fence – the motion of the source leads to a periodical light signal. As the detector collects the signals from the whole of the grating, that is, from all image points that pass the sensor's field of vision, A, the output signal, $S(t)$, can mathematically be described as the cross-correlation signal between the two-dimensional intensity distribution of the surface image, $i(x, y)$, and the aperture function of the grating, $a(x, y)$. When sensor and surface move with relative velocity v along the x-axis, $S(t)$ is given as

$$S(t) = \iint_A a(x, y) \cdot i(x - vt, y) \, dx \, dy \qquad (2.10)$$

In contrast to the time-domain correlation sensors, this is a sensor that "calculates" the cross-correlation of two spatial functions. It is interesting to note that this calculation is not carried out in a microcontroller unit, but in the sensor's optical system and, thus, at the speed of light.

Due to the convolution theorem, the transformation of Eq. (2.10) into frequency space yields the signal power spectrum as the product of the power spectrum of the grating and the power spectrum of the image intensity distribution. As the latter is the result of a transformation of the height profile into brightness values, it is also a measure for the spatial frequency spectrum of the surface. Thus, the power spectrum of the grating can be identified as the transfer function of the sensor: it transforms the spatial frequency spectrum of the surface, moving with velocity v, into a periodical signal with fundamental frequency f_0, which is given as (g is the grating constant and M the scale factor of the imaging lens):

$$f_o = \frac{M \cdot v}{g} \qquad (2.11)$$

When a pulse train is formed out of the periodic signal, every pulse can be considered as one length increment $\Delta l = g/M$. Then, the measured path length, l, can be calculated from the pulse count, N, to

$$l = N \cdot \Delta l = N \cdot \frac{g}{M} \qquad (2.12)$$

In this sense, the sensor can also be counted to the displacement sensors. As the detector collects a periodical intensity signal from every minute structure of the surface, and as all of these signal trains carry the same frequency, the integration over the whole image field will yield a sensor signal with this frequency as well. Due to the random phases of these signal trains, however, there will also be constant and low-frequency components in the integrated signal (Eq. 2.10), which will contribute to the system noise and limit the dynamic range of the sensor. The solution is to employ two complementary gratings in one optical setup and to

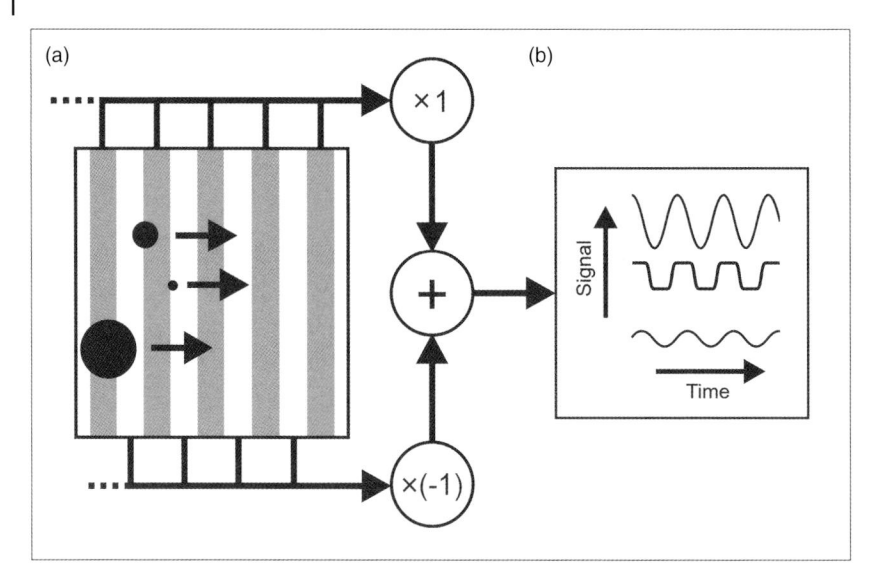

Figure 2.12 Cross-correlation sensor for velocity measurements. (a) Schematic setup, (b) signals resulting from structural components of different sizes.

collect their signals with two photodetectors. Their alternating signal components will have opposite phases, whereas their static and low-frequency components will be identical. Subtracting the signals from the two gratings will thus eliminate the latter components and double the alternating, velocity-dependent ones. The two complementary gratings are usually not realized by two separated beam paths, but by splitting the image plane into stripes that run perpendicular to the direction of movement. Upon connecting every second stripe, the arrangement resembles two combs, with the teeth of one comb filling the gaps of the other (Figure 2.12). The signals from these complementary gratings are of opposite phases, and by multiplying them with +1 or −1, respectively, and adding them up in an electronic circuit, the result is the desired alternating signal with velocity-dependent frequency.

As can be seen from Figure 2.12, the size of the surface structures determines the form of the output signal: when the structure size matches the size of the grating gaps, the signal becomes almost sinusoidal. When the structure is larger than the gaps, the modulation amplitude decreases, and the static signal components increase, and when the structure is smaller than the gaps, the signal becomes almost rectangular. In any case, however, the frequency will only be determined by the grating constant and the relative velocity between sensor and surface. The distribution of signal amplitudes, the signal power spectrum, will, on the other hand, depend on the surface structure. This explanation is equivalent to that of the sensor providing the transfer function between spatial frequency spectrum and signal power spectrum, but it may be a little more illustrative. The sensor principle is favorably applied to stochastically structured materials, like road or rail

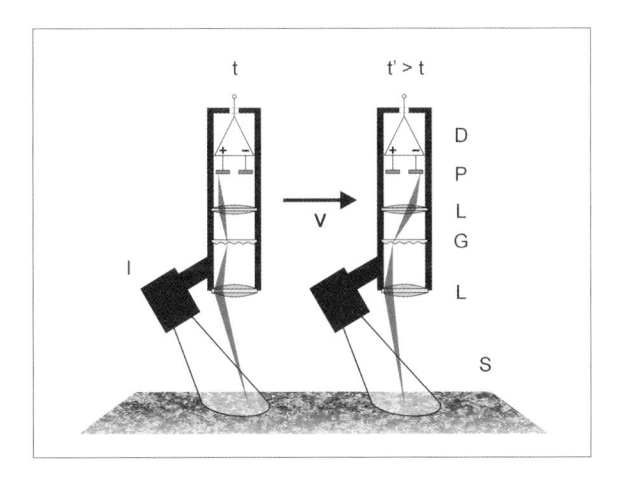

Figure 2.13 Schematic setup of an optical-grating cross-correlation sensor. I: illumination, S: surface, L: lens, G: grating, P: photodetectors, D: differential amplifier.

head surfaces, and industrial materials like paper, carpet, or sheet metal. In case of periodically structured surfaces, however, the sensor signal contains beat frequencies due to the interaction of surface structure and grating. In this case, the analysis of the sensor signal becomes extremely difficult so that periodical surfaces are not suited for this measurement principle.

The schematic setup of Figure 2.12 can be realized in different ways. An optical way is to employ prism gratings (Figure 2.13) that act as periodical beamsplitters. When the surface is illuminated by an incoherent light source, either by a halogen lamp or an LED, an image point that is imaged onto the grating will alternately be projected onto the two photodetectors. The second lens system acts as a field lens by which the integration over the entire field of view is achieved.

The grating constants of these sensors are around $g = 500\,\mu m$. With a scale factor of $M = 0.5$, a velocity of $v = 100\,m/s$ will lead to a signal frequency of $f_0 = 100\,kHz$. From Shannon's theorem, the sampling frequency of the sensor electronics must exceed 200 Hz; hence quite fast circuits are required for the digitization of the signal. Subsequently, the signal is filtered and the periodic signal is transformed into pulses that can then be counted and evaluated digitally.

As the grating can also be designed to capture two orthogonal velocity components [20], not only longitudinal, but also transverse vehicle dynamics can be measured. Gratings with four flanks, that is, with kinks exactly in the middle of each flank, require four photodetectors, and the alternating components from each of them will have relative phase shifts of 90°. As in the case of incremental encoders, these quadrature signals allow the determination of the direction of movement [21]. Diffraction will not be an issue in these sensors unless the grating constants become significantly smaller than 100 μm. For these small grating constants, the described sensor setup is not suited, although smaller grating constants may be necessary for surfaces with very fine textures. Also, the grating constants cannot be adapted to different surfaces, and the sensors have a relatively large number of optical components [22].

An alternative sensor concept combines grating, field lens, and detectors into one single optoelectronic element: a CCD or a CMOS detector array. This direct embodiment of the measurement principle in Figure 2.12 not only reduces the adjustment steps to a minimum, but also reduces the overall size of the optical setup. Also, it provides grating constants in the micrometer range. One option of such a setup is that the grating constants can be varied by grouping two or more detector elements. There are also two-dimensional detector arrays that can measure two-dimensional velocities and others that yield quadrature signals for direction detection [22].

A typical sensor of this type is employed for measuring the dynamics of road vehicles without the influence of wheel slip. It illuminates the road surface with a halogen lamp, covers a velocity range from 0.5 km/h to 250 km/h and measures path lengths with an uncertainty of 0.2% [23]. Longitudinal and transverse velocity components are used to calculate the slip angle, the angle between the vehicle's longitudinal axis and the actual velocity vector, in a range between −40° and +40°, and with an angle resolution of 0.1°. As an additional challenge, these sensors are subject to continuous changes in their ride heights. For usual imaging optics, this would result in a continuous change of the scale factor of the lens, M, in Eq. (2.11). To keep it constant, it is therefore necessary to equip the sensor with a telecentric lens. Its telecentric stop will also ensure a large depth of field so that the sensor described above has a working distance of 350 mm and a working range of ±100 mm around this point (Table 2.7).

Apart from the detector arrays, one can also employ fast camera modules. As these are available commercially, this solution is cheaper, but also less flexible. One sensor of the described type is designed for industrial applications, particularly for the measurement of velocities in the production of sheet metal [24]. It comes with an LED illumination and covers a velocity range of 1–3000 m/min with a measurement uncertainty of ±0.05%. Its working range covers up to ±30 mm around the working distance of 300 mm.

Table 2.7 Technical specifications of a correlation velocity sensor [23].

Property	Value
Velocity measurement range	0.5–250 km/h
Velocity measurement uncertainty	<±0.2%
Angle measurement range	−40° to +40°
Angle measurement uncertainty	<±0.1°
Working distance and range	(350 ± 100) mm
Path length resolution	1000 pulses/m
Operating temperature	−25°C to +50°C

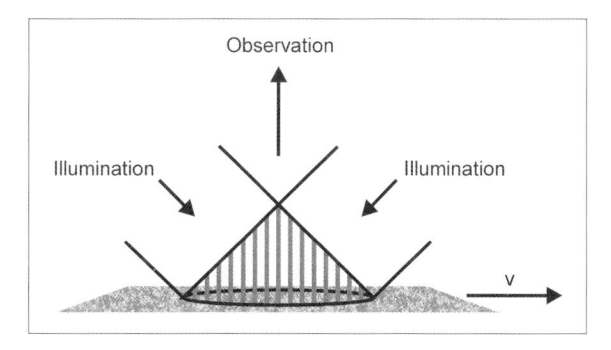

Figure 2.14 Illumination and observation geometry of a velocity sensor with a grating of laser light.

Table 2.8 Technical specifications of a laser-Doppler sensor [25].

Property	Value
Measurement range	0–±3600 m/min
Measurement uncertainty	<±0.05 %
Working distance and range	(240 ± 10) mm
Laser output power	15 mW (780 nm)
Path length resolution	Up to 10.000 pulses/m

The optical grating can also be realized with a special illumination geometry. When two laser beams cross, parallel planes of high intensity will form in the intersection zone due to interference effects (Figure 2.14).

As the interference planes are strictly parallel, the grating constant will not change when the distance between sensor and surface varies. Thus, the telecentricity of the optical setup is already realized in the illumination geometry. For the optical detection path, there is only need for an imaging lens that collects the light from the entire field onto the detector. Here, the constant and low-frequency signal components must be eliminated by electronic filtering. A sensor working on this principle measures velocities between 0 m/min and 3600 m/min with a typical measurement uncertainty of 0.05 %. Here, the working distance is, for example, 240 mm with a working range of ±10 mm, see also Table 2.8 [25].

It is interesting to note that this type of sensor is also denoted as laser Doppler sensor and works similar to a principle applied for measuring the velocities of particles in gas and liquid flows in up to three dimensions, laser Doppler anemometry (LDA). It is a dual-beam technique in which two laser beams interfere at their intersection point (Figure 2.15). When a particle crosses the beam intersection, it scatters light into the direction of the detector. Because of the opposite directions of the beams with respect to the particle, the backscattered light carries opposite Doppler shifts, and the detector receives their beat frequency.

In the interference model, an interference pattern of parallel planes forms in the intersection volume in the same way as described above. This interference pattern can be understood as a grating of light with grating constant, g, which is determined by the wavelength of the laser light, λ, and by the angle of intersection, θ:

$$g = \frac{\lambda}{2\sin(\theta/2)} \qquad (2.13)$$

Thus, light scattered from particles that cross the interference pattern with velocity v will result in an amplitude-modulated signal with frequency $f = v/g$ (Figure 2.15). Combining this with Eq. (2.13) results in

$$v = \frac{\lambda f}{2\sin(\theta/2)} \qquad (2.14)$$

The scattered light intensity is usually measured in the direction of maximum intensity. The forward direction is generally most suited, but the backward direction has the advantage to have laser and detector in the same housing and is thus preferred in industrial solutions. In general, the angle of intersection must match the size of the particles in the flow. For a given grating constant, g, the modulation amplitude will decrease with increasing particle size. For a detailed discussion of

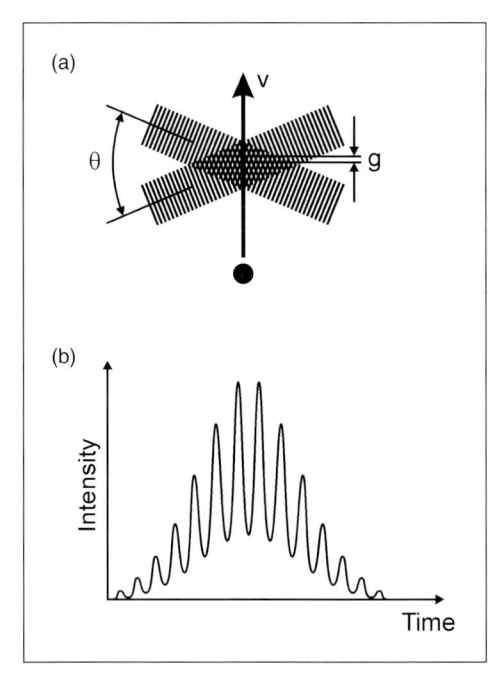

Figure 2.15 Laser Doppler anemometry. (a) A particle crossing the intersection of two laser beams, (b) detector signal.

the LDA principle, see Ref. [26]. As an example, an LDA setup is capable of measuring the orientation of human blood capillary loops close to the skin surface *in situ* [27]. Here, the flow velocities of the blood were in the range of some 0.1 mm/s.

Laser Doppler setups that can also determine the direction of movement utilize AOMs as additional optical elements. As they produce a frequency modulation of the output beams, the motion of a particle will lead to a Doppler shift of the modulation frequency, and the sign of this shift will be direction dependent. Due to the expensive components, however, this setup will be comparably costly.

It is an inherent prerequisite for the proper function of a laser Doppler sensor that the wavefronts must be perfectly planar so that the distance between the lines of the interference pattern is constant. Otherwise, the measured frequency would depend on the distance between object and sensor. The other way round, this may also be employed for distance measurement: a recently described setup [28] utilizes two crossed pairs of laser beams with different wavelengths. One of the pairs yields a light grating that is diverging with distance, and the other is converging. This has the effect that an object point that changes its velocity will yield frequency changes of the same sign in both beam pairs, while a change in distance yields frequency changes into opposite directions. With this technique, measurement uncertainties of around 0.1% become possible.

For a detailed theoretical survey about the basics of noncontact velocimetry based on correlation techniques, see Ref. [29].

6.2.4
Angular Velocity

As mentioned in the last section, a sensor that measures the two-dimensional speed of a vehicle over ground is applied to measure its slip angle. To this end, one needs an additional sensor that determines the yaw rate, $d\Psi/dt$, and the coordinates of the velocity sensor must be known in the vehicle's center-of-mass system. This coordinate system is usually defined with the $+x$-axis pointing from the center-of-mass to the front of the car, the $+y$-axis to its left side, and the z-axis pointing upward. When $v_{\text{Sensor},x}$ and $v_{\text{Sensor},y}$ denote the longitudinal and transverse velocity components, respectively, and a_x and a_y the sensor's x and y coordinates, the slip angle, β, becomes

$$\beta = \frac{v_{\text{Sensor},y} - \frac{d\Psi}{dt} \cdot a_x}{v_{\text{Sensor},x} - \frac{d\Psi}{dt} \cdot a_y} \tag{2.15}$$

The yaw rate, $d\Psi/dt$, has the dimension of an angular velocity. A sensor to measure the yaw rate is usually based on an optical gyroscope. Fiber gyroscopes are actually Sagnac interferometers formed by a number of fiber loops (Figure 2.16): polarized light is coupled into both ends of an optical fiber and split up into two counterpropagating components by means of a beamsplitter. Upon leaving

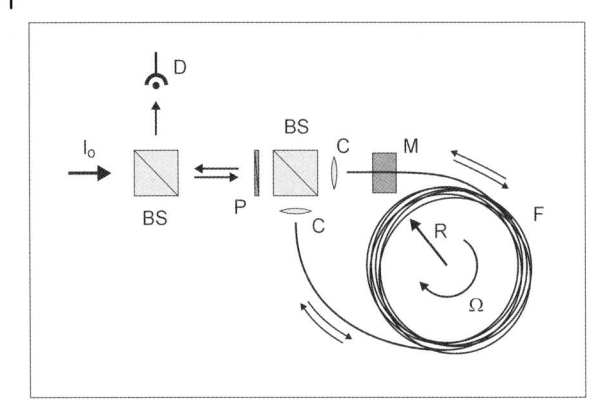

Figure 2.16 Fiber gyroscope. I_0: incident intensity, BS: beamsplitter, P: polarizer, C: fiber coupler, M: frequency modulation, F: fiber loop, R: radius of fiber loop, Ω: angular velocity, D: photodetector.

the fiber loops again, they interfere in the beamsplitter. The polarizer is necessary to compensate any polarizing effects, for example, due to induced birefringence in the fibers. The outgoing beam is finally coupled into a photodetector with a beamsplitter cube.

When the fiber loops of radius R rotate with an angular velocity of Ω, a path difference for the two counterpropagating beams develops and, thus, they experience a phase shift $\Delta\Phi$ of [30]:

$$\Delta\Phi = \frac{8\pi NA}{\lambda c} \Omega \tag{2.16}$$

where $A = 2\pi R^2$ is the area covered by the fiber loops and N the number of fiber loops. This phase shift, the well-known Sagnac effect, leads to a change in the intensity received by the photodetector and is a direct measure for the angular velocity of the gyroscope.

The detected signal, I_D, is (I_{D0} represents signal for $\Omega = 0$):

$$I_D = \frac{I_{D0}}{2}(1 + \cos \Delta\Phi) \tag{2.17}$$

This signal is symmetrical, that is, phase shifts of different signs result in signals of identical signs. Also, the small phase shifts to be expected from fiber gyroscopes will only lead to small signal changes due to the cosine function. When a constant phase shift of $\pi/2$ is added to $\Delta\Phi$, one gets

$$I_D = \frac{I_{D0}}{2}(1 + \cos(\Delta\Phi + \pi/2)) = \frac{I_{D0}}{2}(1 - \sin \Delta\Phi) \tag{2.18}$$

As the sine function crosses zero at $\Delta\Phi = 0$, both signs become possible, and the larger signal changes with small phase shifts increase the sensitivity of the gyroscope significantly. Due to the mathematical properties of the sine function, the detector signal is approximately linear for small values of $\Delta\Phi$. An even higher sensitivity can be reached by adding a periodical phase shift with amplitude β and angular frequency ω to Eq. (2.17):

$$I_D = \frac{I_{D0}}{2}(1+\cos(\Delta\Phi+\beta\sin\omega t)) \qquad (2.19)$$

The signal output will also be periodic with the modulation frequency, and it can be shown [30] that the output components modulated with the odd multiples of ω are proportional to $\sin\Delta\Phi$. Thus, the detector signal can be evaluated with a lock-in amplifier whose output can be adjusted to yield the amplitude of the first harmonic, that is, the signal component modulated with ω.

As an example, a commercial gyroscope [31] can measure angular velocities of $\pm 375°/s$. Its linearity is better than 500 ppm for angular velocities below $\pm 150°/s$, its full measurement error is better than 1500 ppm (Table 2.9). The mean time between failures (MTBF) for this sensor is given as more than 55 000 hours. Sensors of this type are generally employed in inertial navigation systems (INS). As described above, one typical application is the measurement of yaw, pitch, and roll angles in vehicle testing applications, but they are also used as INS in ships or planes.

For extremely accurate measurements, for example, of the Earth's angular velocity, a similar instrument, the laser gyroscope, is employed [32]. It consists of a laser resonator, a planar arrangement of several mirrors in which two laser beams circulate in opposite directions. One possible technical solution is a square of vacuum tubes, filled with a mixture of helium and neon. In one of the tubes, there is a glow discharge between two high-voltage electrodes that confine the amplifying medium of a HeNe laser. The physical description of this gyroscope is similar to that of the fiber gyroscope, but with the additional condition that in a ring resonator, a whole number of wavelengths, n, has to fit into the resonator length, L. A small change ΔL in the effective resonator length leads to a frequency shift $\Delta \nu$ between the counterpropagating beams of

$$\Delta \nu = \nu \frac{\Delta L}{L} = \frac{c\Delta L}{\lambda L} \qquad (2.20)$$

Table 2.9 Technical specifications of a fiber gyroscope [31].

Property	Value
Measurement range	$0°/s$ to $\pm 375°/s$
Linearity	<500 ppm (below $150°/s$, room temperature), <1000 ppm (above $150°/s$, room temperature)
Angle random walk	$4°/h\,Hz^{1/2}$
Total measurement error	<1500 ppm
Operating temperature	$-50°C$ to $+85°C$

Hence, the frequencies of the counterpropagating beams correspond to neighboring longitudinal modes of the laser. The phase shift is obtained in analogy to Eq. (2.16), from which the frequency shift results as

$$\Delta v = \frac{4A}{\lambda L} \Omega \tag{2.21}$$

This frequency shift between the counterpropagating beams will lead to a beat frequency at the output of the ring laser. In general, laser gyroscopes have higher linearities than fiber gyroscopes, and they are less sensitive to temperature changes and show smaller angle drifts. Both systems are, in general, quite expensive and are therefore only used in high-tech applications.

In principle, rotating interferometers and laser resonators must both be treated on the basis of Einstein's theory of general relativity, but the result is the same as for the nonrelativistic treatment [33]. Recently, a ring laser with an enclosed area of $16\,m^2$ has been set up at the Research Site Wettzell of the German Federal Office of Cartography and Geodesy at Kötzting, Germany. The project is still underway, but the final target is to measure changes of the Earth's rotation in the range of one part per billion (1 ppb) relative to the length of day [34].

6.3
Strain

Resistive strain gages are state-of-the-art in strain measurement. They utilize the strain-induced change of a conductor's electrical resistance due to the change of its geometry and its specific resistance. These strain gages, however, have their limitations, particularly when external electromagnetic interferences disturb their sensor signals.

An optical alternative is the fiber Bragg grating (FBG) sensor. This sensor consists of an optical fiber with a periodical inner structure. It is manufactured by placing an ultraviolet-sensitive, germanium-doped quartz fiber into the intersection zone of two excimer laser beams that create an interference pattern with equidistant lines (see above). This can already be done while the fiber is drawn, that is, during the manufacturing process. The result is a zone in the fiber core in which the index of refraction changes periodically. When light from a broadband source, for example, from an SLED, is coupled into this FBG of grating constant g and with an effective index of refraction n_{eff} across the FBG, the spectrum of the reflected light will show a sharp peak at the Bragg wavelength, λ_B, which is given according to

$$2n_{eff} g = \lambda_B \tag{2.22}$$

This wavelength will be missing in the transmitted light spectrum (Figure 2.17). According to Eq. (2.22), the deformation of a fiber subject to some external strain will result in a change of the grating constant and, thus, to a change in the reflected wavelength.

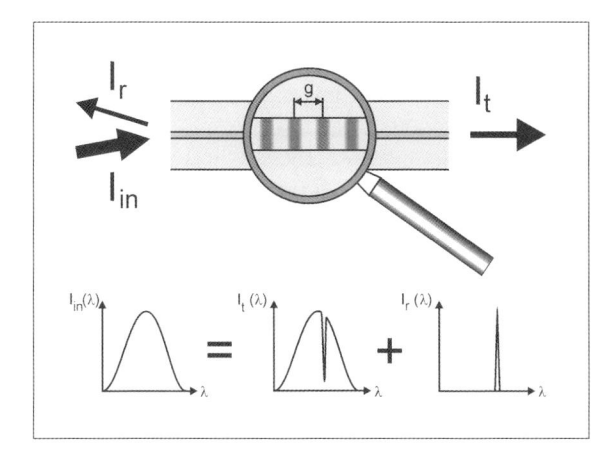

Figure 2.17 Fiber–Bragg grating. g: grating constant, I_{in}: incident intensity, I_r: reflected intensity, I_t: transmitted intensity. The arrow thicknesses indicate the amount of the respective intensities, the diagrams show the spectra of incident, reflected, and transmitted light.

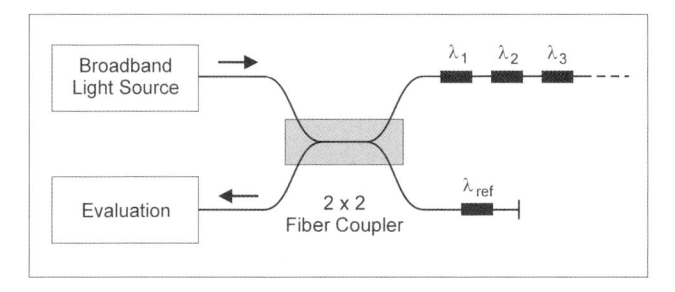

Figure 2.18 Fiber–Bragg grating sensor network. λ_1, λ_2, λ_3: Bragg wavelengths for strain measurement, λ_{ref}: Bragg wavelength of reference grating.

It is a general principle in interferometry that the bandwidth of the reflected spectrum is inversely proportional to the number of interfering rays, that is, to the number of FBG lines. Although the modulation of the index of refraction is only between 10^{-4} and 10^{-3}, the large number of grating periods will result in reflectances of more than 99% [35].

The back-reflected light is coupled out of the FBG fiber with a fiber coupler. As light source and detector can then be housed in the same sensor unit, there is only need for one connection between the FBG and the evaluation circuit. Thus, the number of connections for FBG strain sensors is only half the number of connections for electrical strain gages – a large advantage in terms of costs for material and assembly. Also, one fiber may contain several optical strain gages with different grating constants. According to Eq. (2.22), each of these has its own Bragg wavelength. When each of these Bragg wavelengths is evaluated for its respective shift, strain values on different positions and on even complicated workpieces can be measured with only one fiber (Figure 2.18). This interrogation principle is denoted by wavelength division multiplexing (WDM) or optical frequency domain reflectometry (OFDR). Another interrogation technique, time division multiplexing (TDM) or optical time domain reflectometry (OTDR), exploits the different times of flight of light reflected back from the different FBGs.

In principle, not only strain but also temperature changes lead to a change of the grating constant due to the thermal expansion of the fiber. This temperature dependence can lead to a change in the strain reading of as high as 800 µm/m at a temperature change of 100 K, which is about eight times the value for a resistive strain gage under the same conditions [36]. In total, the relative change of the Bragg wavelength with strain ε (in µm/m) and with temperature can be expressed by [37]

$$\frac{\Delta\lambda_B}{\lambda_{B0}} = (1-p_{\text{eff}})\varepsilon + [\alpha_n + (1-p_{\text{eff}})\alpha_S]\Delta T \tag{2.23}$$

with λ_{B0} is the Bragg wavelength at temperature T_0 and at zero strain, p_{eff} the effective photoelastic coefficient, α_n the thermo-optical coefficient of the FBG, α_S the thermoelastic coefficient of the structure to which the FBG is applied, and ΔT is the temperature change relative to T_0. In real-life situations, it is therefore hard to tell how much of the FBG's wavelength shift is contributed by external strain, and how much of it is caused by a temperature change. For exact measurements, it is therefore necessary to employ a reference FBG for temperature compensation. This FBG has no mechanical contact to the workpiece under investigation and is therefore not subject to strain forces. It only measures the temperature in the vicinity of the optical strain gage (Figure 2.18).

Apart from the temperature dependence, there are some other characteristic differences between resistive strain gages and FBG strain sensors: One important property of a strain gage is the so-called k factor, which is defined by

$$\frac{\Delta x}{x} = k\cdot\varepsilon \tag{2.24}$$

Here, x is the measurement property of the strain gage, so that x equals the electrical resistance R for resistive gages, and $x = \lambda$ for FBGs. For FBGs, k is about 0.8, which is 40% of the value for a typical resistive strain gage [36]. This means that FBGs are generally more "rigid" than resistive strain gages. Also, the temperature range over which FBGs can be employed reaches from 0 °C to 80 °C, while their electrical counterparts cover a range from −200 °C to +200 °C. One of the large advantages of the FBGs is that their fatigue behavior is superior to that of resistive strain gages: While the former sustain 10^7 load cycle changes at strain amplitudes of 3000 µm/m with virtually no zero drift, the latter sustain the same number of load cycle changes at strain amplitudes of only 1000 µm/m with a zero drift of less than 30 µm/m [36].

The generally accepted technique to interrogate FBG strain gages is to measure the wavelength shift with a unit that both contains the light source, usually emitting in the so-called C band between 1530 and 1565 nm, and the spectrum analyzer. The specifications of this analyzer are very demanding: According to Eq. (2.22), a

Table 2.10 Technical specifications of an FBG interrogator [38].

Property	Value
Number of optical channels	4–16
Wavelength range	1510–1590 nm
Wavelength accuracy	1 pm
Wavelength stability	1 pm
Wavelength repeatability	0.5 pm (1 Hz), 0.2 pm (0.1 Hz)
Operating temperature	0 °C to +50 °C

strain resolution of $1\,\mu m/m$ ($= 10^{-6}$) requires a wavelength resolution in the picometer range. Commercially available devices have a wavelength accuracy and stability of 1 pm [38] (Table 2.10).

In principle, the spectral evaluation of the FBG output is the most robust way to interrogate an FBG because it makes the evaluation independent of intensity variations of the light source. All evaluation principles based on intensity measurements suffer from lower precision and stability. For a detailed description of these and all other aspects of optical strain measurements, the interested reader is referred to Ref. [39].

Optical strain sensors can easily be employed for the structural monitoring of buildings. As an example, several FBG sensors were integrated into the concrete structure of a newly built bridge in Wetzlar, Germany. The purpose of this project was to collect experience with the handling of these sensors and also to create a public demonstrator for the possibilities of optical sensors [40]. Apart from the optical and interrogation issues, a very crucial point when measuring strains with FBGs is the coupling of the fiber core to the workpiece. To this end, the cladding must be removed and the core must be attached to the workpiece with a "patch" that ensures a rigid coupling of the strain forces into the FBG.

A fiber-optical sensor that does not rely on the deformation of a rigid structure is the extrinsic fiber Fabry–Pérot interferometer (EFPI) sensor [41] (Figure 2.19). A glass capillary with a length of some millimeters and a diameter of some tenths

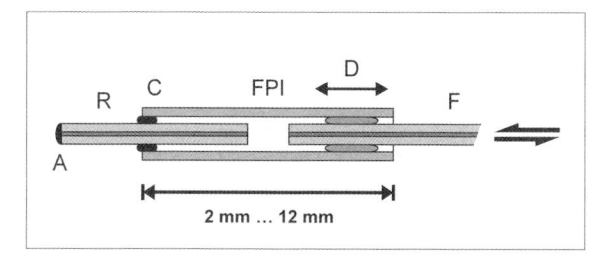

Figure 2.19 EFPI sensor, schematic setup. F: input fiber, FPI: Fabry–Pérot interferometer, C: glass capillary, R: reflecting fiber, A: absorber, D: direction of strain.

Table 2.11 Technical specifications of EFPI sensor [41].

Property	Value
Strain measurement range	−2% to +2.5%
Strain resolution	<10^{-5} (10^{-7} for short-time measurements)
Sensor length	10–25 mm
Sensor diameter	0.5 mm
Diameter of connecting fiber	0.6–1 mm
Force threshold	150–200 µN

of a millimeter holds two fiber ends. The end through which the light enters, and which also guides the light back to the evaluation unit, can slide inside the capillary. The other fiber end is fixed and holds an absorber outside the capillary. Inside the capillary, the fiber end faces form a Fabry–Pérot interferometer (FPI). These interferometers are generally employed as resonators in lasers, and their output spectrum shows equidistant interference fringes.

The frequencies of the fringes is described by Eq. (1.7) and is a function of the resonator length, that is, of the spacing between the two fiber ends of the EFPI sensor. When strain is applied to the sensor, the spectral fringe positions will change with the resonator length according to

$$\frac{\Delta v_q}{v_q} = \frac{\Delta L}{L} \qquad (2.25)$$

When the photodetector in the evaluation receives two subsequent consecutive fringes when strain is applied to the sensor, this will correspond to a change in the resonator length of one-half of the wavelength that is coupled into the EFPI sensor.

The sensor described in Ref. [41] measures strain for applied forces of more than 150–200 µN, and strain resolutions in the µm/m range can be achieved (Table 2.11). Thus, the measurement performance of an EFPI sensor is comparable to that of an FBG sensor, but its setup is more complicated and a sensor network cannot be constructed as conveniently as with FBGs.

6.4
Temperature

In fact, precise temperature measurements require good thermal contact between the thermometer and the reservoir whose temperature is to be determined. Whenever, on the other hand, the location of the measurement is hard to access, when electromagnetic interferences do not allow the use of electronic sensors or when the temperature is so high that contacting thermometers would simply be destroyed, optical or, in general, noncontact temperature measurement becomes

Table 2.12 Technical specifications of a thermal FBG probe [42].

Property	Value
Measurement range (standard)	−40 °C to + 120 °C
Thermo-optical coefficient	(9.9 ± 1.7) pm/K
Measurement uncertainty (calibrated)	±0.5 °C (long-term), ±0.2 °C (short-term)
Response time	0.2–8.5 s (depending on probe type)
Center wavelength	1462–1618 nm

necessary. With noncontact techniques, however, temperature can only indirectly be deduced from other, temperature-induced effects.

The first example for an optical thermometer is the FBG described in the last section. As indicated by Eq. (2.20), the wavelength reflected by an FBG changes linearly with absolute temperature, T. A typical thermal FBG probe [42] has a thermo-optical coefficient, α_n (Eq. 2.20), of 9.9 pm/K, so that a wavelength resolution of 1 pm allows a temperature resolution of about 0.1 K. This probe covers a standard temperature range from −40 °C to 120 °C (Table 2.12). WDM or TDM setups allow the distributed temperature measurement with only one fiber containing several FBGs.

The most direct approach to noncontact temperature measurements is to make use of the properties of thermal radiation. As described in Part One of this book, the law of Stephan and Boltzmann, Eq. (1.1), allows to calculate the absolute temperature of an object from the total radiant power that it emits. A well-known technique to measure the temperature of a glowing body is pyrometry. Pyrometres are the only practical sensors for temperatures of more than 1800 °C, and they are suited for a temperature range from 650 °C to 3000 °C.

Figure 2.20 shows the schematic setup of a visual pyrometer. The glowing object is imaged onto a glowing filament. Through an eyepiece, both the image of the surface and the object can be observed in the same image plane. When both

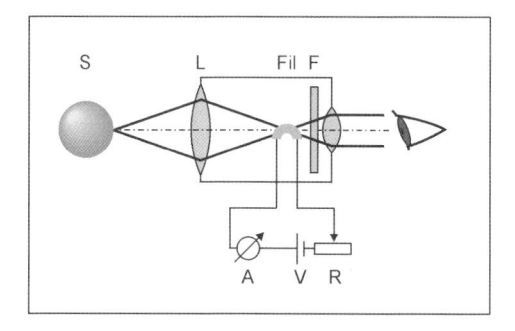

Figure 2.20 Visual pyrometer, schematic setup. S: glowing surface, L: lens, Fil: glowing filament, F: red filter, A: current meter, V: voltage source, R: adjustable resistor.

Table 2.13 Technical specifications of an electronic pyrometer [43].

Property	Value
Measurement range	−32 °C to +600 °C
Measurement uncertainty (at 23 °C)	1 K, or at least 1% of measurement value
Spot size/distance	30:1
Aperture	15 mm
Emissivity	0.2–1.0
Operating temperature	0 °C to +55 °C

radiant intensities match, the filament has virtually no contrast against the glowing object, and as the electrical power applied to the filament is known, the temperature can be calculated with ease.

Although the basic principle is easy to understand, the problem here is that most glowing surfaces are not ideal blackbodies, and their emissivities are smaller than 1. Before the temperature can be measured with high precision, one has to determine the emissivity of the object of interest. This can either be done by heating up a sample of the material and measuring its temperature with a contacting thermometer, or by bringing the material into thermal contact with a reference material of known emissivity. When the temperature of the reference is known, the emissivity of the material can be determined. It may also be possible to drill a hole into the glowing body and take this as an ideal blackbody with emissivity $\varepsilon = 1$ and finally reference the object's emissivity to it. After such a calibration procedure, temperatures can be measured with accuracies of around 1%.

Electronic pyrometers can be realized with thermopile detectors. One of these devices [43] covers a temperature range from −32 °C to 600 °C. Its measurement uncertainty reaches 1 °C or at least 1% of the measured temperature value (Table 2.13). Handheld pyrometers usually have a pilot laser with which the operator can exactly target the object whose temperature shall be measured. There are not only industrial, but also clinical applications for this type of thermometers: They measure the body temperatures of babies, little children or other persons for whom the usual oral or rectal measurement is too uncomfortable or who may be too uncooperative. Also, these traditional techniques may even communicate infections between different persons. Electronic thermometers basically determine the temperature of the eardrum. It shares blood with the hypothalamus so that changes in the body core temperature are measured with virtually no time delay [44].

Thermography is a technique for the measurement of two-dimensional temperature distributions. Thermographic cameras are designed very similar to cameras for the visible spectrum, but their optics have to be optimized for long wavelengths. They are usually made of polyethylene, silicon, germanium, or sodium chloride for wavelengths between 8 µm and 14 µm. The type of imaging sensor depends on the wavelength range. Table 2.14 shows an overview of the relevant imaging arrays.

Table 2.14 Detectors for thermographic cameras.

Wavelength range (μm)	Detector
0.8–1.1	Silicon
1–2	InGaAs, PbS
2–5	Bolometer
1–5	InSb
8–14	GaAs

Table 2.15 Technical specifications of an uncooled thermographic camera [45].

Property	Value
Measurement range	−20 °C to +500 °C (optional: +250 °C to +1200 °C)
Measurement uncertainty	2 K, or at least 2 % of measurement value
Thermal sensitivity (at 30 °C)	0.08 K
Resolution microbolometer array	320 × 240 pixels
Frame rate	25 Hz
Spectral range	7.5–13 μm
Emissivity	0.1–1.0
Operating temperature	−15 °C to +45 °C

Detectors are often cooled to reduce noise from thermal background, even down to the temperature of liquid nitrogen. An uncooled thermographic camera built around a microbolometer array with 320 × 240 pixels [45] works in two optional temperature ranges between −20 °C and 500 °C or between 250 °C and 1200 °C, respectively, and with a measurement uncertainty of ±2 % of the measurement value, or ±2 °C, whatever value is higher (Table 2.15). As with pyrometers, the cameras require a correction for the emissivity of the target object.

One typical application for thermographic cameras is the detection of thermal losses of buildings. In industry, they are used for the quality and safety control of electrical or electronic devices because every component that fails will show a significant increase in temperature.

6.5
Species Determination and Concentration

There are several effects by which particle concentrations influence light. Probably the best-known effect is the absorption of light, widely utilized by a variety of spectrometric techniques. Another is optical activity that rotates the plane of polarization of the incident light. And finally, scattering and diffraction are the effects upon which the detection of aerosoles is based.

6.5.1
Spectrometry

Light that is transmitted through a liquid, gaseous, or solid sample will suffer from extinction. The transmitted intensity, $I(d)$, depends on the length of the path that the light travels through the medium, d, and on the wavelength-dependent absorption coefficient of the material, $\alpha(\lambda)$. With I_0 as the incident intensity, this is described by Lambert–Beer's law:

$$I(d) = I_0 \cdot e^{-\alpha(\lambda)d} \tag{2.26}$$

For liquids, the absorption coefficient can be expressed as the product of the so-called molar absorptivity, $\varepsilon(\lambda)$, and the concentration, c, of the absorbers:

$$\alpha_{\text{liquids}}(\lambda) = \varepsilon(\lambda) \cdot c \tag{2.27}$$

The absorption coefficient of gases is usually expressed as the product of absorption cross section, $\sigma(\lambda)$, and the number density, N, of the absorbers:

$$\alpha_{\text{gases}}(\lambda) = \sigma(\lambda) \cdot N \tag{2.28}$$

From Eq. (2.26), the absorbance A is defined as

$$A = -\ln\left(\frac{I(d)}{I_0}\right) \tag{2.29}$$

The absorption spectrum of the sample, that is, the spectral characteristics of the absorbance, is described by the spectral characteristics of the absorption coefficient. Every species that is contained in the sample will contribute to this spectrum with its specific spectral fingerprint: Atoms show line spectra with well-separated lines, molecules show a multitude of broad and equidistant vibration lines (which may even have some inner structures due to rotational transitions) for each electronic transition, and solid objects may have a totally irregular absorption spectrum. There are two ways to exploit this effect, and they differ by the result that is desired by the analysis. When the spectrum is recorded as a whole, that is, with a continuum light source and a polychromator, it is possible to determine both the different species contained in the sample and their respective contents. If only one species is to be determined at a time, the light transmitted through the sample passes a monochromator tuned to a wavelength that is specific for the respective species (Figure 2.21). When this species is a chemical element,

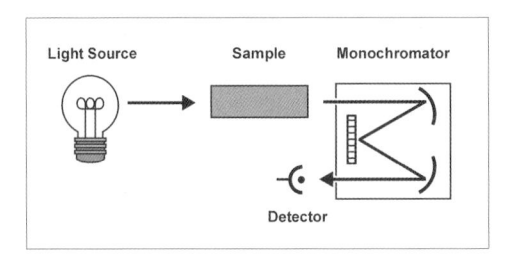

Figure 2.21 Schematic setup of a spectrometer.

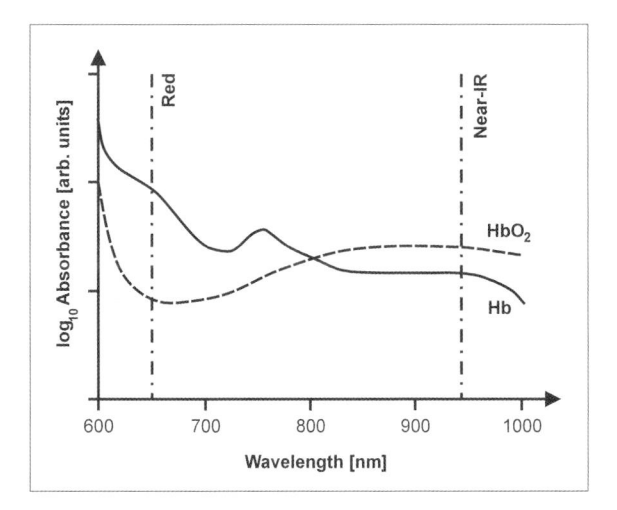

Figure 2.22 Spectral absorbance of oxygenated hemoglobin (HbO$_2$) and reduced hemoglobin (Hb). Vertical lines represent spectral positions of detection light sources.

it is also possible to use a line source, usually a so-called hollow-cathode lamp, that emits the element-specific wavelengths directly. Continuum sources, however, have the great advantage of emitting enough intensity even in spectral regions where no line sources are available or where they do not emit with sufficient intensity.

One example for a spectrometric technique is pulse oximetry, a method employed, for example, in emergency departments or intensive care units for the determination of the degree of oxygen saturation of a patient's blood. This property is usually expressed in terms of the functional hemoglobin saturation, SaO$_2$. It is calculated with the concentrations, c, of oxygenated hemoglobin (HbO$_2$) and reduced hemoglobin (Hb):

$$\text{Functional SaO}_2 = \frac{c_{HbO_2}}{c_{HbO_2} + c_{Hb}} 100\% \tag{2.30}$$

Usually, the value for healthy persons is between 96% and 100%. The working principle of a pulse oximeter is based on different absorbances of HbO$_2$ and Hb in the red and near-infrared spectrum (Figure 2.22, adapted from [46]). At around 650 nm, the absorbance of Hb is by about one order of magnitude larger than that of HbO$_2$, and at around 940 nm, it is significantly smaller. A pulse oximeter contains LED light sources with these two wavelengths in its sensor head and subsequently calculates the functional SaO$_2$ in its evaluation unit.

The emitter–detector unit is incorporated into a clip that can easily be attached to a patient's finger, ear, or nose. A cable connects the clip with an evaluation and display unit. Besides the functional hemoglobin saturation, a time-resolved measurement also yields the pulse rate. As only arterial blood shows this pulse rate, its oxygenation corresponds to the AC component of the signal while that of venous blood corresponds to its DC component. The evaluation unit determines both

Table 2.16 Technical specifications of a pulse oxymeter [47].

Property	Value
Functional SaO_2 measurement range	70%–100% (calibrated)
Functional SaO_2 resolution	1%
Pulse measurement range	25–240/min
Pulse resolution	1/min
Operating temperature	0 °C to +50 °C

absorbance components ($AC(\lambda)$, $DC(\lambda)$) for both wavelengths and calculates the ratio R [46]

$$R = \frac{AC(660 \text{ nm})/DC(660 \text{ nm})}{AC(940 \text{ nm})/DC(940 \text{ nm})} \quad (2.31)$$

The oximeter's memory contains a lookup table in which the R values are related to the corresponding SaO_2 values. These value pairs were determined from experiments in which test persons inhaled air mixtures of decreasing oxygen content. As these experiments are only carried out with oxygen contents that do not subject the test persons to medical problems, SaO_2 values of 80% or less can only be extrapolated and are thus measured with reduced accuracies. Table 2.16 shows the specifications of a commercial pulse oxymeter [47].

Pulse oximeters show systematic measurement errors whenever the light path contains elements that change the absorption properties of the medium. In case of a finger clip, artificial finger nails and nail varnish (except when it is red or purple) lead to reflections and to additional attenuation of the probe light. Also, circulatory disorders may even lead to the impossibility to apply pulse oximetry. If the patient moves unintentionally or shivers, wrong measurement values may result. Finally, it should be noted that although a pulse oximeter measures the functional hemoglobin saturation, it gives no information about the total oxygen content of the blood.

Another example for a spectrometric technique is the quality monitoring of drinking water. Chemical elements contained in water samples are determined with spectrometers based on atomic absorption spectrometry (AAS): The water sample is brought into a graphite furnace and dried. All nonvolatile components are then atomized at temperatures of typically 2000 K. An alternative to this electrothermal atomization (ETA) is the atomization of the sample in an air/acetylene flame. In both cases, a setup with a continuum source and a polychromator is suited for simultaneous multielement determination. The wavelengths selected for the determination are mostly in the ultraviolet part of the spectrum. In any case, however, it is necessary to calibrate the device with samples of different known element concentrations.

The detection limit (LOD) of an analytical technique is defined as the analyte concentration, given in µg/l, that corresponds to three times the standard deviation

Table 2.17 LODs for various elements with continuum-source AAS in an air–acetylene flame.

Element	Wavelength (nm)	LOD (µg/l)
Ag	328.1	1.2
Cd	228.8	0.8
Cr	357.9	1.8
Ir	208.9	760.0
Li	670.8	0.12
Pb	217.0	10.0
Pd	247.6	4.0
Zn	213.9	1.4

of a blank sample. Table 2.17 lists up examples of LODs for various elements with the respective absorption lines [48]. Interested readers can find a more detailed description of AAS in Ref. [49].

As monochromators or polychromators require an entrance slit that defines the spectral resolution, spectrometric techniques that directly measure the absorption spectrum cannot make use of the entire amount of light that has passed the sample. Thus, the signal-to-noise ratio may be poor for certain samples. This is one motivation that led to the invention of Fourier-transform infrared (FTIR) spectroscopy. Its basic setup is the same as for a Michelson interferometer: light from an infrared source is split up by a beamsplitter into two interferometer arms. There, mirrors reflect the two partial beams back to the beamsplitter where they are combined back to one beam that is finally directed through the sample to the detector (Figure 2.23).

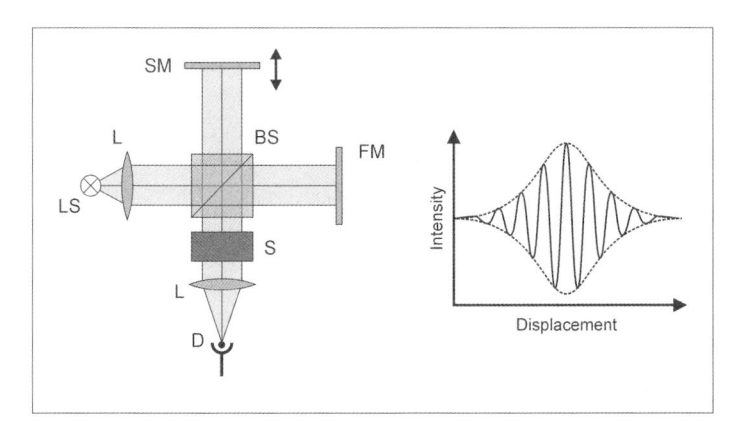

Figure 2.23 Fourier-transform infrared (FTIR) spectroscopy, schematic setup. LS: light source, L: lens, BS: beamsplitter, SM: scanning mirror, FM: fixed mirror, S: sample, D: detector.

For a light source that emits a spectrum $I(v')$ with a bandwidth Δv, the intensity received by the detector is a function of the displacement x of the scanning mirror from the position where both interferometer arms are of equal length. With the wavenumber $v' = 1/\lambda$, the intensity received by the detector is [50]

$$I(x) = \int_{-\infty}^{+\infty} I(v')W(v')\cos(2\pi v'x)\,dv' \qquad (2.32)$$

Here, $W(v')$ is a window function that equals 1 across the bandwidth of the light source and zero otherwise. By replacing the cosine term with Eulers formula (the additional sine term does not contribute to the integral) and by applying the inverse transformation in order to find an expression for the spectrum, one gets

$$I(v') = \int_{-\infty}^{+\infty} I(x)W(v')e^{i2\pi v'x}\,dx \qquad (2.33)$$

Thus, the spectrum $I(v')$ results from the Fourier transform of Eq. (2.32), calculated from interferogram $I(x)$. Apart from the advantage of working with higher intensity levels, FTIR provides the measurement of the whole spectrum in one shot by recording a time-domain signal. The time required for one measurement is in the range of a few seconds or even less. Also, the setup can be miniaturized so as to enable the realization of portable devices. More details on the application of FTIR spectrometers can be found, for example, in Ref. [51].

6.5.2
Polarimetry

Polarimetry is based on the effect of optical activity, or natural circular birefringence: when an optically active sample is illuminated with linearly polarized light, the plane of polarization is rotated. This is usually detected in an arrangement where the sample is placed between two crossed polarizers so that any change in the state of polarization leads to an increase in the intensity behind the polarizers.

The effect can be exploited with all substances whose crystal or molecular structures show so-called chiral asymmetry, characterized by the absence of any mirror or inversion symmetry. One example is a molecule with at least one so-called asymmetric carbon atom, that is, a carbon atom bonded to four different atoms or groups. This is the case for the molecule of dextrose, or D-glucose ($C_6H_{12}O_6$): A solution of dextrose in water rotates linearly polarized light to the right (clockwise), when viewed toward the light source.

The rotation angle, α, is proportional to the concentration c of the substance and to the length l of the light path within the substance. With the specific rotation α^*, the rotation angle is (Biot's law)

$$\alpha = \alpha^* \cdot c \cdot l \qquad (2.34)$$

Table 2.18 Specific rotation, α^*, of an aqueous solution of sucrose at a temperature of 20°C for light sources of different wavelengths (elements given in brackets).

Wavelength (nm)	$\alpha^*(°/10\,\text{cm g cm}^{-3})$	Wavelength (nm)	$\alpha^*(°/10\,\text{cm g cm}^{-3})$
670.8 (Li)	50.51	480.0 (Cd)	103.07
643.8 (Cd)	55.04	467.8 (Cd)	109.69
636.2 (Zn)	56.51	435.8 (Hg)	128.49
589.3 (Na)	66.45	419.1 (Fe)	140.0
546.1 (Hg)	78.16	388.9 (Fe)	166.7
515.3 (Cu)	88.68	382.6 (Fe)	173.1

The specific rotation depends on the substance that is investigated, on the employed wavelength, and on the temperature of the sample. As an example, the wavelength dependence of the specific rotation, the so-called rotatory dispersion, of sucrose is given for wavelengths between 408 nm and 644 nm by the formula [52]

$$\alpha(\lambda) = \left(a + b\left(\frac{\lambda}{\mu m}\right)^2 + c\left(\frac{\lambda}{\mu m}\right)^{-4} + d\left(\frac{\lambda}{\mu m}\right)^{-8} \right) \cdot \alpha(0.5462271\,\mu m) \qquad (2.35)$$

with $a = -0.0017982$, $b = 0.2765318$, $c = 0.00655736$, and $d = 0.0000103825$. Thus, the rotation angle decreases with increasing wavelength. Equation (2.35) is in accordance with the recommendations of the International Commission for Uniform Methods of Sugar Analysis (ICUMSA) for wavelengths between 540 nm and 633 nm [53]. Table 2.18 shows the specific rotation of sucrose for different wavelengths at a temperature of 20°C [54].

Besides its dependence on the temperature, θ, the specific rotation of a solution will also show a slight dependence on the concentration, c [55]. The length of the light path in the substance is usually given in multiples of 10 cm because this is the traditional length of the sample tubes. The specific rotation of a substance can be determined by measuring the rotation angles for different concentrations. Then, Eq. (2.34) can be applied to determine the concentration of this substance in an unknown solution. This is done with a polarimeter (Figure 2.24).

Because of rotatory dispersion, the wavelength dependence of specific rotation, the light source of a polarimeter must be monochromatic. The usual configuration is a line source in combination with a filter to select one of the emission lines. In most cases, polarimeters utilize yellow light with a wavelength of around 589.3 nm, the center-of-mass wavelength of the sodium D lines. An alternative is the green light of the mercury isotope ^{198}Hg at 546.2 nm. While the polarizer defines the plane of polarization of the incident light, the analyzer is set to full extinction with an empty sample tube: polarizer and analyzer are "crossed." A photodetector behind the analyzer monitors the transmitted intensity, and when an optically active sample is inserted into the polarimeter, the optical rotation, α, will lead to a detector signal that is proportional to intensity $I(\alpha)$. When I_0 is the intensity transmitted by the open polarizer–analyzer pair, one gets

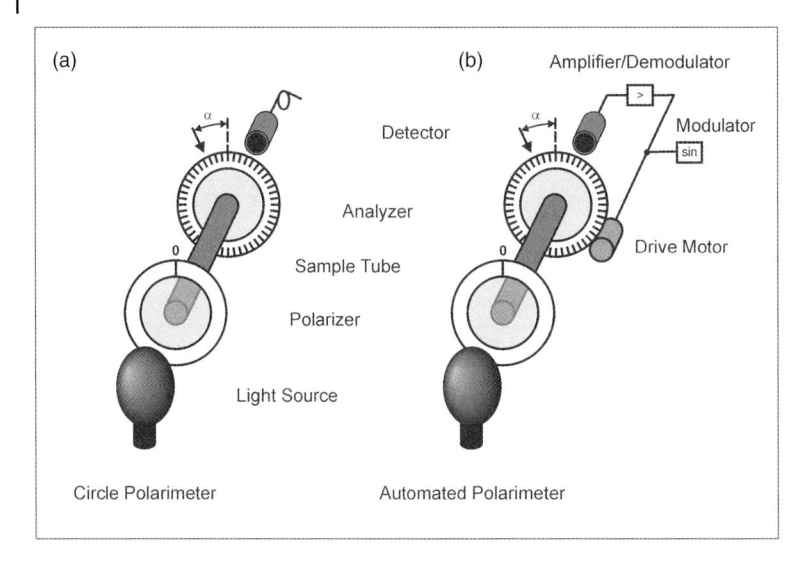

Figure 2.24 Schematic setup of a polarimeter. (a) Circle polarimeter, (b) automated circle polarimeter. α: rotation angle. The device in (b) shows a mechanic modulator of the analyzes.

$$I(\alpha) = I_0 \sin^2 \alpha \tag{2.36}$$

This simple setup is usually called a circle polarimeter. The optical rotation is compensated by rotating the analyzer into the opposite direction until the detector signal becomes zero again. A rotational encoder then measures the angle of rotation. For this kind of polarimeter, the angle resolution is in the 0.01° range. The detector signal can also be fed into a feedback loop to adjust the intensity back to zero automatically. Here, an inherent problem of Eq. (2.36) is that both the signal and its change rate become very small during the angle compensation; thus, the point of full compensation is hard to determine and thus a phase-sensitive detection is used. The necessary modulation is introduced via an additional, periodically changing angle of rotation by means of the Faraday effect in a glass rod between polarizer and analyzer. These automated polarimeters reach angle resolutions of 0.001° (Table 2.19, [56]).

It is interesting to note that in visual polarimeters, that is, in polarimeters that use the human eye as a detector, a very old working principle is still in use. The setup of these half-shadow polarimeters is shown in Figure 2.25: It is basically that of the circle polarimeter in Figure 2.24, but the polarizer is segmented into two halves. No matter if this is done with a half-wave plate behind the polarizer (Laurent type) or with an additional half-field polarizer (Lippich type), the effect is that the polarizing axes of the two segments differ by the so-called half-shadow angle, δ. When the analyzer rotates, extinction of the two half-fields will not occur simultaneously, but with an angle difference that corresponds to the half-shadow

Table 2.19 Technical specifications of an automated polarimeter [56].

Property	Value
Measurement range	0° to ±360° (0°Z to ±259°Z)
Resolution	0.001° (0.01°Z)
Measurement uncertainty	±0.005° (±0.02°Z)
Measurement wavelength	589 nm (Na D lines)
Sample temperature, measurement range	0°C–99°C
Sample temperature, measurement uncertainty	±0.1°C

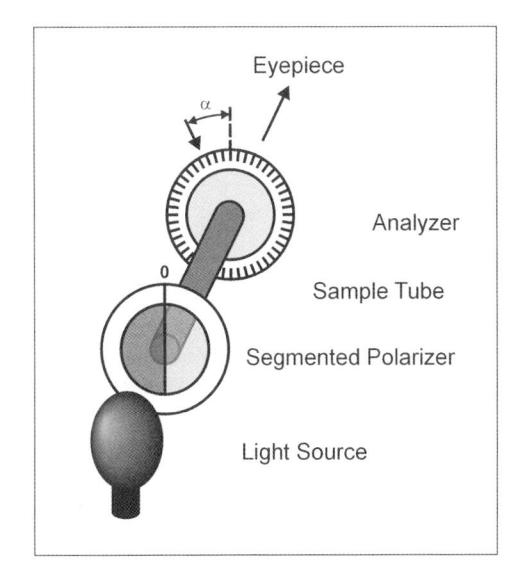

Figure 2.25 Schematic setup of a half-shadow polarimeter. α: rotation angle.

angle (Figure 2.26). If, on the other hand, the polarizing axis of the analyzer is rotated from one extinction position by half the half-shadow angle toward the other extinction position, the intensities of the two half-fields will be identical. This is the position to which the polarimeter must be adjusted to compensate the rotation angle of the sample. The idea behind this principle is that the human eye can distinguish between intensity differences of as low as 2% and that the angle compensation can be done with an angular resolution of 0.1%.

In commercial half-shadow polarimeters, the half-shadow angle is 8° for yellow sodium light ($\lambda \approx 589$ nm) and 9.5° for green mercury light ($\lambda \approx 546$ nm). Low-cost versions may also employ tungsten halogen lamps with color-glass filters. The device [57] has a measurement range of ±180°, reaches an angle resolution of

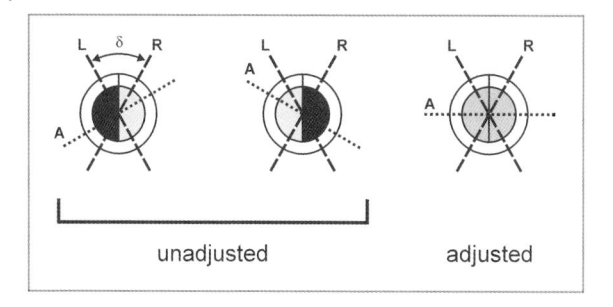

Figure 2.26 Working principle of a half-shadow polarimeter. L/R: left/right polarizer polarizing axis, respectively, A: analyzer polarizing axis, δ: half-shadow angle.

0.1%, and a measurement uncertainty of ±0.1%. When the angle is measured with a vernier scale, and with a little experience, this uncertainty may be further reduced down to ±0.05%.

For further information about the different types of polarimeters, see Ref. [58]. The devices are used in clinical applications for the measurement of sugar concentrations in blood and urine, and also for the quality control in industrial applications, for example, for monitoring the sugar content in the beverage industry. It is interesting to note that in these industries, precise polarimeters, also known as "saccharimeters," measure the sugar concentration by determining the optical rotation in °Z according to the International Sugar Scale (ISS). 100°Z corresponds to the rotation angle of the so-called normal sugar solution: 26.000 g of sucrose are weighed in air under normal conditions of temperature, pressure, and relative humidity (20°C, 1013 mbar, 50%) by using brass weights without buoyancy correction. This mass of sucrose is dissolved in pure water at a temperature of 4°C to result in 100 ml of solution. At 20°C, the concentration of this solution is 237.018 g/kg, and its density is 1.097 639 g/cm^3 [59]. According to ICUMSA regulations, the rotation angle corresponding to 100°Z at 20°C, for a sample tube length of 200 mm and for a vacuum wavelength of 546.2271 nm (green mercury light) amounts to 40.777°.

6.5.3
Ellipsometry

Ellipsometry is a technique to investigate physical properties of thin films like their thickness, refractive index, and/or homogeneity. The technique is based on the physical fact that linearly polarized light, incident onto a dielectric surface under some angle Φ, will be reflected as elliptically polarized light. As a starting point, the incident beam is described in terms of its polarization components perpendicular to or parallel to its plane of incidence (Figure 2.27), usually denoted the s- and p-polarized components, respectively. The field components E_s and E_p can thus be written as (E_{s0}, E_{p0} are amplitudes of s- and p-components, respectively, k the wavenumber, and ω is the angular frequency)

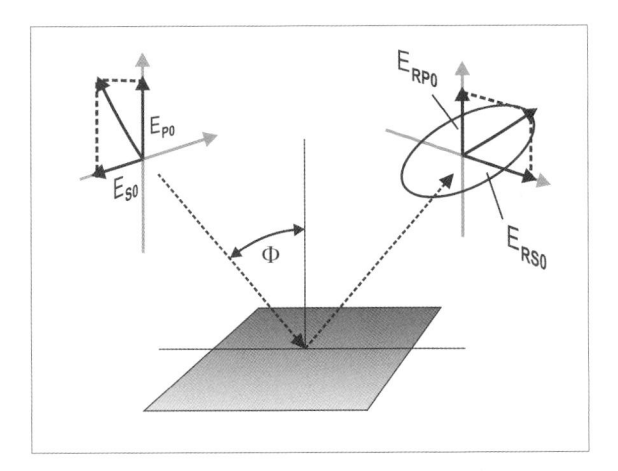

Figure 2.27 Ellipsometry, schematic working principle. E_{S0}, E_{P0}: amplitudes of incident s- and p-polarized light components. E_{RS0}, E_{RP0}: amplitudes of reflected s- and p-polarized light components. Φ: angle of incidence.

$$E_S(\vec{r}, t) = E_{S0} e^{i(\vec{k}\cdot\vec{r}-\omega t)}; \; E_P(\vec{r}, t) = E_{P0} e^{i(\vec{k}\cdot\vec{r}-\omega t)} \tag{2.37}$$

Linear polarization implies that the two components propagate with no relative phase shift. The components of the reflected beam are then represented by

$$E_{RS}(\vec{r}, t) = E_{RS0} e^{i(\vec{k}\cdot\vec{r}-\omega t)} \tag{2.38}$$

$$E_{RP}(\vec{r}, t) = E_{RP0} e^{i(\vec{k}\cdot\vec{r}-\omega t+\Delta)} \tag{2.39}$$

Thus, the s- and p-polarized components of the reflected beam have different amplitudes, and they experience a phase shift. The result is elliptically polarized light whose electric field vector rotates around its wave vector while it propagates (Figure 2.27). Finally, the parameter that is evaluated in ellipsometry is

$$\rho = \frac{E_{RP}}{E_{RS}} = \tan\Psi \cdot e^{i\Delta} \tag{2.40}$$

Here, Ψ is the angle between the major axis of the elliptically polarized light and the plane of incidence. From Eq. (2.40), a model-based evaluation allows to determine the desired properties of the sample, and the input data to this analysis may be the precalculated with the reflection coefficients obtained with the Fresnel equations [3].

Ellipsometry thus requires the measurement of the state of polarization of the reflected light. This is usually done by rotating an analyzer in the reflected light path and recording the light intensity that reaches the detector as a function of the rotating angle (rotating analyzer ellipsometry, RAE). There are also setups in which the polarizer is rotated (rotating polarizer ellipsometry, RPE).

As the optical parameters of the sample are wavelength dependent, the simplest solution is to employ a laser light source. Also, ellipsometers may contain addi-

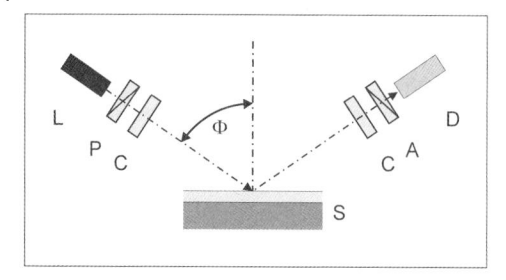

Figure 2.28 Schematic setup of an ellipsometer. L: laser light source, P: polarizer, C: compensator (optional), S: sample, A: analyzer, D: detector. Φ: angle of incidence.

Table 2.20 Specifications of a commercial ellipsometer [60].

Property	Value
Wavelength range (max.)	245–1690 nm, 670 wavelengths
Angle of incidence	Approx. 65°
Data acquisition rate	<6 s per surface point
Sample area (max.)	1.1 m × 1.5 m

tional polarization-optical components, for example, retardation plates as compensators (Figure 2.28).

Sophisticated ellipsometers also feature the analysis on different wavelengths, so-called spectroscopic ellipsometers. These devices employ continuum light sources, and the reflected spectrum is analyzed with a monochromator. Other models detect the reflected light with a long-distance objective and an imaging detector, and are capable of measuring film thicknesses and refractive indices over a larger sample area in one measurement. Table 2.20 shows the specifications of a high-speed spectroscopic ellipsometer [60].

6.5.4
Refractometry

A refractometer is an instrument that measures the refractive index of a sample, mostly a liquid substance. Pure water, for example, has a refractive index of 1.33, and every substance that is diluted in it will lead to a change in its refractive index. Thus when the dependence of the refractive index on the concentration is known for a certain substance, a refractometer can be calibrated to yield concentration values. As this dependence is different for different substances, each refractometer must be calibrated for the substance whose concentration it is intended to measure.

All refractometers make use of a particular case of Snellius' law of refraction. When light travels from a medium with a high refractive index to one with a low

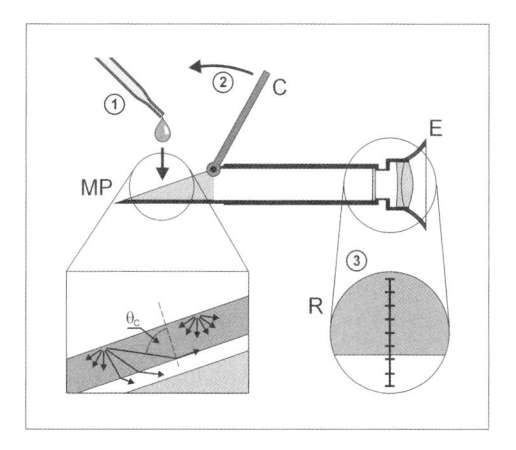

Figure 2.29 Schematic setup of a refractometer. C: cover plate, MP: measuring prism, E: eyepiece, R: reticule. Measurement steps: 1: Put a drop of water onto the measuring prism, 2: close the cover plate, 3: read the result from the eyepiece reticule.

refractive index, its angle of refraction will be larger than the angle of incidence. As described above (Section 6.1.2), there is a critical angle of incidence, θ_C, above which light cannot enter the second medium: it is totally reflected from the interface. When the first medium has a refractive index of $n_R \approx 1$ (vacuum or air), Eq. (2.1) is modified to

$$\sin\theta_C \approx n_i^{-1} \qquad (2.37)$$

For water with $n_i = 1.33$, the critical angle is about 49°. Refractometers make use of this principle by means of the setup shown in Figure 2.29, which shows a simple, visual handheld refractometer.

Here, one drop of the substance under investigation is put onto the measuring prism. The cover plate, a ground-glass plate, is closed and the liquid is dispersed over the entire prism face. The cover plate scatters incident light to enter the sample at arbitrary angles, and all angles of incidence that are larger than the critical angle will be totally internally reflected inside the liquid. It follows from Eq. (2.37) that the refractive index of the prism must be larger than that of the sample for the refractometer to work properly.

The critical angle leads to a sharp horizontal border between a dark and a bright field in the field of view of the refractometer's eyepiece when looking through it against a light source. A reticule with one or more scales inside the eyepiece is calibrated to allow a direct reading of the measurement value.

Depending on the desired result, the refractometer scales are calibrated in different units. A unit that is widely used in the fruit or beverage industry is degree Brix (°Brix). As all substances that can be investigated with a refractometer contain sugar, °Brix is a unit for their density and sugar content: a liquid has 1°Brix when it has the same density as a solution of 1 g of sucrose in 100 g of a sucrose/water solution. The more °Brix a solution has, the sweeter it tastes, and the less perishable it will be. Thus, the measurement value helps to estimate its quality.

As an example, a value between 6°Brix and 8°Brix is a sign of low quality for fruit juice, highest qualities can be assumed for values of around 20°Brix. For vegetables with their low sugar contents, highest qualities are indicated by values between 10°Brix and 15°Brix.

Another possible refractometer scale is the Oechsle scale. It measures the density of grape must, an important quantity to estimate its quality. The more degree Oechsle (°Oe) it has, the higher the alcoholic strength of the fermented wine will be. In Germany, an average vintage reaches between 70 and 80°Oe, which corresponds to an alcoholic strength between 10% and 11%.

As the refractive index of the liquid is a function of temperature, one must either specify the temperature during the measurement, or the refractometer must compensate any thermal influences. In handheld devices, this is usually done by means of additions glass wedges mounted on bimetal strips. When the temperature changes, the wedge will shift the border between darkness and brightness with respect to the fixed reticle in the field of view. A commercial instrument [61] is capable of compensating temperature influences between 10°C and 30°C. It has a measurement range from 0°Brix to 32°Brix and a measurement uncertainty of ±0.2%. A digitization of such a refractometer can be done in a very straightforward manner by capturing the field of view with an imaging detector and detecting the border between the dark and bright fields. Then, the temperature compensation can also be carried out digitally. One of these refractometers [62] reaches a measurement range from 0°Brix to 90°Brix and a measurement uncertainty of ±0.1%.

Laboratory refractometers are so-called Abbe refractometers, of which both visual and digital versions exist. Their basic working principle is same as for handheld devices, but they have a built-in light source and may compensate temperature influences by stabilizing the temperature of the optical elements, usually at a temperature of 20°C. To account for the dispersion of the measurement prism, the measurement is done with yellow sodium light at about 589 nm, the wavelength of the sodium D lines. This is why the refractive index measured with a refractometer is usually symbolized by

$$n_D^{20} = n\,(589\text{ nm}, 20°C) \qquad (2.38)$$

A digital Abbe refractometer always comes with the possibility to transfer data to and from a personal computer. One of these [63] can measure values of up to 95°Brix in a compensated temperature range between 0°C and 90°C, and with a measurement uncertainty of 0.1°Brix. As the corresponding measurement range for the refractive index is 1.3000–1.7200, its measurement uncertainty is 0.0002 (Table 2.21).

6.5.5
Particle Density and Particle Number

As described above, a smoke detector can be understood as a special type of light barrier. When the scattered light intensity reaches a preadjusted threshold, an alarm is triggered and indicates the presence of smoke (Figure 2.5).

Table 2.21 Specifications of a digital Abbe refractometer [63].

Property	Value
Measurement range sugar scale	0–95 ° Brix
Resolution sugar scale	0.1 ° Brix
Measurement uncertainty sugar scale	0.1 ° Brix
Measurement range refractive index	1.3000–1.7200 (589 nm)
Resolution refractive index	0.0001 (589 nm)
Measurement uncertainty refractive index	0.0002 (589 nm)

The physical principle behind this detector is the Tyndall effect. It describes the scattering of light by particles of a size that is comparable with the wavelength. This is also described by the Mie scattering of electromagnetic waves by spherical particles. The intensity of the scattered light varies with the scattering angle due to interferences between the scattered light waves. In a measurement geometry with a fixed angle between (infrared) emitter and photodetector, the intensity of the scattered light is proportional to the mass concentration in the measurement volume. As the intensity signal will also depend on the optical properties of the aerosol (smoke or dust) particles, the sensor has to be calibrated. This is usually done by comparing the detector reading with the result of a gravimetric measurement, that is, by weighing the aerosol mass deposited onto a filter.

Figure 2.30 shows the schematic setup of the measuring cell, a so-called scattering photometer, a sensor that exploits the described principle for the measurement of respirable dust concentrations [64]. It is built around an infrared laser diode with a wavelength of 940 nm. The photodetector is arranged at a viewing angle of 70° with respect to the optical axis of the laser. Opaque light stops and light traps prevent the detection of direct light from the light source. For the technical specifications of the sensor, see Table 2.22.

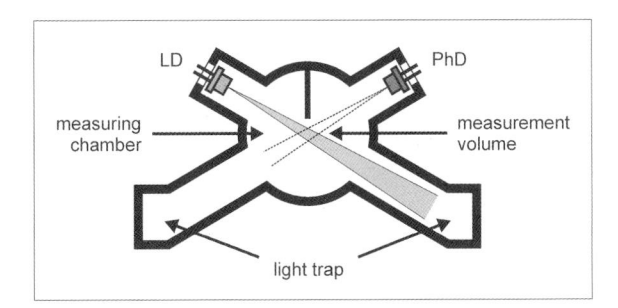

Figure 2.30 Measuring chamber of a respirable dust sensor. Measurement volume: intersection area of light source and detector. LD: laser diode, PhD: photodiode.

The whole sensor is portable and was originally developed for the measurement of respirable dust in the mining industry. Its purpose is the protection of miners from excessive dust concentrations that are likely to cause silicosis, a very serious

Table 2.22 Specifications of a scattering-photometer respirable dust sensor [64] (concentration values for calibration with DEHS particles (monodisperse, diameter $d = 1\,\mu m$).

Property	Value
Measurement range	0.00–99.99 mg/m^3
Linearity	1%
Temperature dependence	5% (0–40 °C)
Limit of detection	10 µg/m^3
Measuring wavelength	940 nm

lung disease. Other applications include the localization of aerosol emission sources, the monitoring of industrial filters, and also the monitoring of workplaces.

The same optical principle is used in a similar instrument that contains a stack of three scattering photometers around a very particular aerodynamic arrangement [65]. A ring-shaped slit inlet allows only inhalable particles with aerodynamic parameters of less than 100 µm to enter the system when the external sampling pump delivers an air flow of 3.1 l/min. With a two-stage virtual impactor, two fractions are separated from this particle collective: the respirable fraction with particle diameters of less than 4 µm is collected in the first stage. A filter in the second stage collects all particles with diameters between 4 µm and 10 µm, which corresponds to the thoracic fraction minus the respirable fraction. The remaining fraction, collected in the third stage, is the so-called extrathoracic fraction with particles of diameters between 10 µm and 100 µm. In each of these three stages, the dust fractions are collected on filters and weighed for a gravimetric analysis of the sampled dust. The results are also used to calibrate the scattering photometers so that a true online measurement of the three dust concentrations becomes possible. The collection of the described fractions makes this device a good technical model of the deposition characteristics of the human lung. The sensor therefore complies with EN 481, the standard that defines the size fractions for the measurement of airborne particles in workplace atmospheres. The sensor reaches a measurement range of 0–250 mg/m^3 and the limit of detection is about 50 µg/m^3 (Table 2.23). Also, a comparative study of six inhalable-aerosol samplers yielded the result

Table 2.23 Specifications of a three-stage scattering-photometer dust sensor [65] (concentration values for calibration with DEHS particles (monodisperse, diameter $d = 1\,\mu m$).

Property	Value
Measurement range	0–250 mg/m^3
Dust fractions	Respirable, thoracic, inhalable
Resolution	50 µg/m^3
Limit of detection	0.1 mg/m^3
Standards	EN 481
Measuring wavelength	780 nm

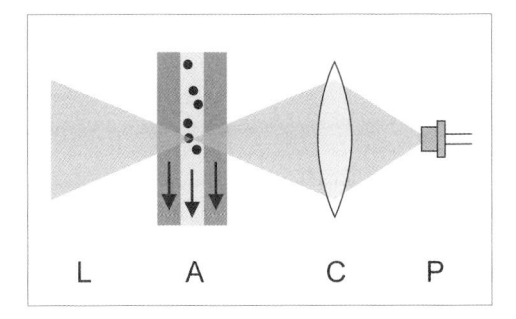

Figure 2.31 Schematic setup of a particle size spectrometer. L: focused laser beam, A: aerosol stream with air sheath, C: collimating optics, P: photodetector.

Table 2.24 Specifications of a particle size spectrometer [67].

Property	Value
Measurement range particle size	0.3–20 µm
Measurement range concentration	$<10^5/cm^3$
Number of channels	Max. 128
Flow rate	3 l/m
Measuring wavelength	632.8 nm

that the described sensor provided a reasonable match with the inhalable convention [66].

In contrast to particle concentrations, particle numbers and sizes are usually determined with laser methods. Figure 2.31 shows a schematic setup: A stream of aerosol particles is aerodynamically prepared so that the particles pass the measurement area one after the other. An air sheath confines the particle stream laterally. A laser beam is focused onto the measurement area and passes the aerosol stream at right angles. As a result, each aerosol particle that passes the laser focus results in an impulse on the photodetector. The number of pulses corresponds to the number of particles, and the pulse height corresponds to the particle size. Such a sensor is therefore called a particle size spectrometer. A typical particle size spectrometer [67] works for particle sizes between 0.3 µm and 20 µm and for concentrations of up to 10^5 particles/cm^3. The pulse height analysis enables the sensor to distinguish between 128 different particle size classes (Table 2.24). A typical application for these devices is air quality monitoring in cleanrooms.

Another application for this sensor principle is incorporated in a so-called disdrometer, a device that can simultaneously measure the particle sizes and veloci-

6 Optical Sensor Concepts

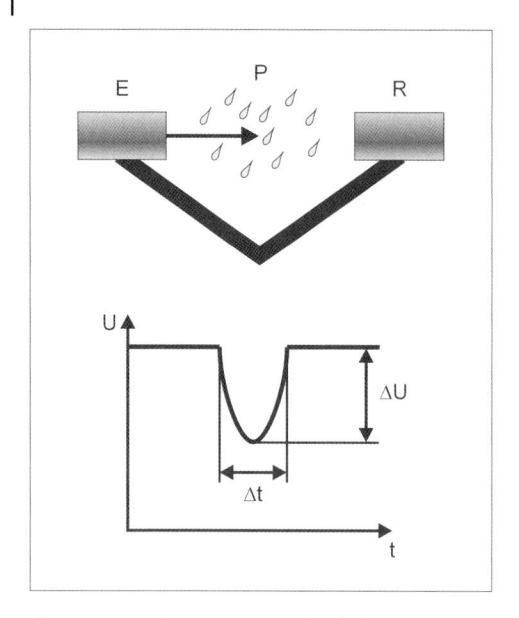

Figure 2.32 Schematic setup of a disdrometer. E: emitter, R: receiving detector, P: precipitation, U: detector output voltage, ΔU: voltage drop due to precipitation, t: time, Δt: time interval of voltage drop.

ties of solid and liquid precipitation. Disdrometers are installed, for example, along motorways or in the vicinity of airport runways, and they provide important data for weather forecasts, for example, for flood warnings. Figure 2.32 shows the principle: A laser light source and a detector unit are mounted on a frame so that their optical axes are aligned. In one practical example [68], the laser beam is formed by a cylindrical lens to leave the emitter as a horizontal sheet of light. Every falling object, in meteorological terms a "hydrometeor," will lead to an interruption of the beam and, thus, to a signal drop in the detector unit. This is a very particular setup of a transmission-type light barrier (see Section 6.1.1).

From the form of the interruption signal behind the detector, the precipitation can be classified. As the detector measures the intensity across the whole of the beam profile, the depth of the signal drop is a measure for the particle size. From the duration of the signal drop, the velocity of the hydrometeors can be determined. Sophisticated signal processing also enables the classification of eight different precipitation types: drizzle, mixed drizzle/rain, rain, mixed rain/snow, snow, snow grains, freezing rain, and hail. To ensure operation in all weather conditions, the device must be heated to protect the device from the build-up of ice. Table 2.25 lists the most important specifications of a commercial device [68].

Table 2.25 Specifications of a commercial disdrometer [68].

Property	Value
Wavelength of light source	650 nm
Precipitation classes	8
Particle size	0.2–25 mm
Particle velocity	0.2–20 m/s
Minimum precipitation intensity	0.001 mm/h
Maximum precipitation intensity	1200 mm/h
Precision of discrimination between classes	>97%
Environmental conditions	(−40 to +70) °C, (0–100) % relative humidity

6.5.6
Fluorescence Detection

When a substance is illuminated with light of a (small) wavelength, its molecules get excited in the sense that electrons "jump" from one electronic state to another with higher energy. These energy states are not quite as sharp as those in atoms, but they are split up into several sublevels that describe different vibration states of the molecule. The excited molecule relaxes into the lowermost vibration level of the electronic state and from there, spontaneous emission leads to a relaxation in a lower-lying electronic state and, thus, to the emission of a light quantum by way of spontaneous emission. The whole process takes only some nanoseconds. This lower state is not necessarily the one that the electron originally came from (Figure 2.33). As a consequence, the molecule emits light during its relaxation, but the energy of the emitted light quanta is generally lower than that of the absorbed ones. In other words, the emitted light wavelength is larger than the wavelength of the exciting light:

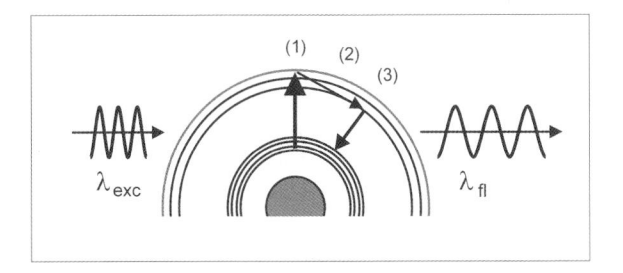

$$\lambda_{fl} > \lambda_{exc} \tag{2.39}$$

Figure 2.33 Fluorescence, schematic of the principle. (1) excitation, (2) relaxation within the vibration levels of the excited state, (3) relaxation by emission of a light quantum. λ_{exc}: excitation wavelength, λ_{fl}: fluorescence wavelength.

This relation is known as Stoke's rule, in memory of the scientist who discovered and systematically investigated this phenomenon in the middle of the 19th century.

There are three different mechanisms that can be exploited for fluorescence investigations:

- Primary (or auto-) fluorescence is an intrinsic property of many biological materials, in particular biological materials, but it can also be achieved by doping with trace elements. One example for this is chlorophyll: when, for example, algae are illuminated with blue light, they fluoresce in the red.

- Secondary (or induced) fluorescence is achieved by staining a nonfluorescent sample with a fluorochrome. As an example, cell nuclei can be stained with acridine orange, and upon exciting the specimen with blue light, it shows green fluorescence.

- Immunofluorescence is a technique in which the fluorochrome couples with antibodies. These antibodies can be designed very specifically for certain biological targets so that they result in an extremely selective staining result.

One fluorochrome that is widely used for immunofluorescence applications is fluorescein isothiocyanate (in short, fluorescein or FITC). This fluorochrome is used to label and track cells in many fluorescence-microscopy applications. Its absorption spectrum reaches from the UV to the green part of the spectrum and is maximal at about 495 nm. The emission spectrum reaches from the blue-green part of the spectrum far into the near-infrared, with its maximum at around 520 nm. Absorption and emission spectra overlap between about 480 nm and 550 nm so that it is necessary to use filter systems for the selection of excitation and fluorescence wavelengths.

Figure 2.34 shows the schematic setup of an optical system that is suitable for fluorescence investigations. The light source, usually a mercury short-arc lamp with sufficient intensity in the short-wavelength regions of the spectrum, sends its light to an excitation filter with which the desired excitation wavelength range is selected. A dichroic beamsplitter reflects the excitation light to an objective with the specimen in its focus. The decisive parameter of this objective is its numerical aperture, which determines the intensity of the excitation light in the specimen plane. The objective thus also serves as a condenser for the light source. The fluorescence light is then picked up by the same objective and passes the dichroic beamsplitter as well. The beamsplitter therefore requires a special coating that reflects, in the example of FITC, blue light and transmits green light. Finally, a barrier filter with high transmission for the fluorescence wavelengths blocks the rest of the excitation light that may still be present in the light path. Excitation filter, dichroic beamsplitter, and barrier filter are a combination that is unique for a particular fluorochrome, and they are usually offered as a so-called filter set. Figure 2.35 shows the schematic transmission curves for an FITC filter set: The excitation filter is a bandpass filter with a bandwidth (FWHM) of about 50 nm. It is an interference filter with very steep transmission edges and some residual ripple across the transmission range due to interference effects of the filter's

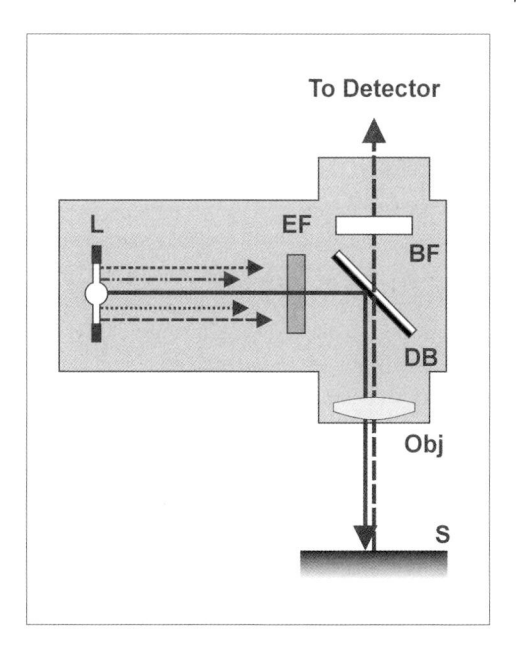

Figure 2.34 Fluorescence detector, schematic setup. L: light source, EF: emission filter, DC: dichroic beamsplitter, Obj: objective, S: specimen, BF: barrier filter.

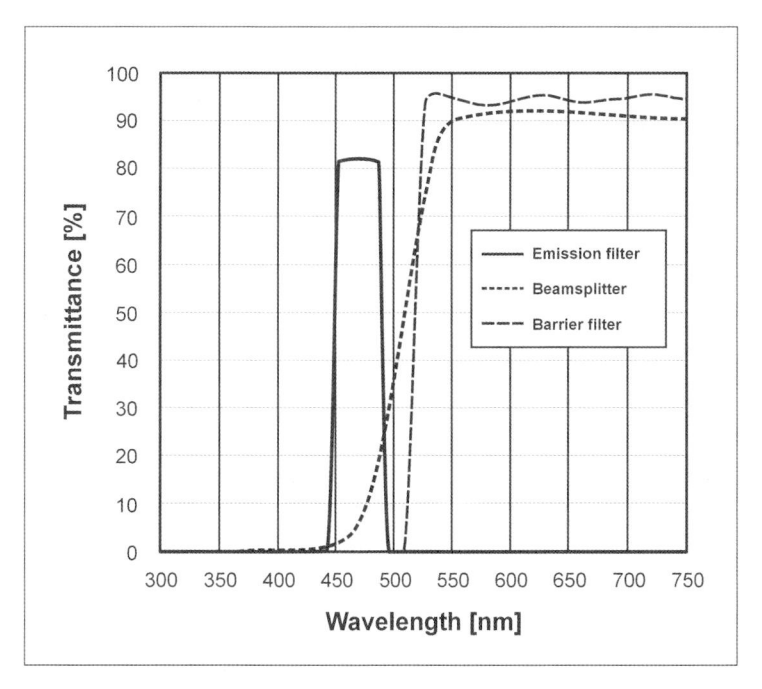

Figure 2.35 Fluorescence filter set for FITC. Transmittance curves for excitation filter, dichroic beamsplitter, and barrier filter.

dielectric layer system. These ripples can also be observed in the transmission ranges of the dichroic beamsplitter and the barrier filter, in this example designed as long-pass filters. The barrier filter also has a steep edge: the transmission jumps from zero to about 95% over a wavelength interval of less than 10 nm. As this edge is well outside the transmission range of the excitation filter, no excitation light will reach the detector.

The fluorescence intensity may be evaluated in two different ways: if the detector is a single photodetector, it will detect the fluorescence intensity, integrated over the objective's entire field of view. A detector signal will then indicate, for example, the presence of cells of a certain type. To avoid any measurement errors, the setup must guarantee that only fluorescence light from the cells that shall be detected reaches the detector. In order to suppress stray light, the objectives must meet very particular specifications: Their inner surfaces are blackened to suppress light scattering, and they also have special edges serving as light traps to prevent disturbing reflections. Also, the edges of the lenses carry black varnish in order to avoid reflections.

The setup of Figure 2.34 can also be incorporated into an imaging system, for example, a microscope. These incident-light fluorescence microscopes have become an extremely powerful diagnostic tool. Here, the detector is an observation tube with eyepieces and also often with a phototube and a camera to document the results. For these imaging applications, the suppression of stray light is a crucial issue because otherwise, it would destroy the image contrast and might thus make diagnoses difficult. Fluorescence microscopes usually carry several filter sets in a turret or filter slide with which the excitation wavelengths for different fluorochromes can very conveniently be chosen. As filter sets are usually quite expensive, and in respect of the other very particular specifications, fluorescence microscopes belong to the most expensive types of microscopes on the market. Due to the potential of marking very specific properties of cells and their components, however, nearly all newly invented microscopic techniques make use of fluorescence, for example, stimulated emission-depletion (STED) microscopy or fluorescence *in-situ* hybridization (FISH), with which even single genes can be stained and investigated.

With the advent of high-power LEDs, fluorescence microscopy becomes affordable even for routine analyses. Most medical practitioners, for example, gynecologists, test specimens for special kinds of bacteria or fungus. As the application is always the same, no versatile and expensive fluorescence microscope is necessary for this target group. Today, application-specific incident-light fluorescence illuminators have become available with which routine or laboratory microscopes can be retrofitted to allow everyday fluorescence diagnoses [69].

6.6
Surface Topography

In a great variety of applications, the surface topography of an object is of great importance. The surface does not only characterize the appearance of an object, but also a large number of its physical and chemical properties. In principle, the surface profile can be recorded with all techniques already described in the section about distance measuring sensors, but as they determine the distance to one single point, the surface must be scanned. However, there are also whole-field techniques with which the profile over the entire field of vision can be recorded in one shot.

In principle, a surface scanner requires a distance measurement sensor and a scanner that moves the sensor across the surface. Some of these techniques make use of interferometric effects and therefore yield extremely highly resolved and precise results. In general, the optical sensor provides high resolution and accuracy for the measurement of the profile height, while the system as a whole has to realize high lateral resolutions and accuracies by employing highly precise guidances, usually equipped with incremental encoders for measuring the positions in the object plane.

6.6.1
Chromatic Confocal Sensors

One well-known example for a high-resolution and highly precise scanning technique is the chromatic-confocal sensor. The term "confocal" denotes that the setup contains a narrow light source, realized either by a narrow stop in front of an extended light source or by a laser. Also, this source is imaged onto a corresponding stop in front of the detector (Figure 2.36): both stops are thus located in

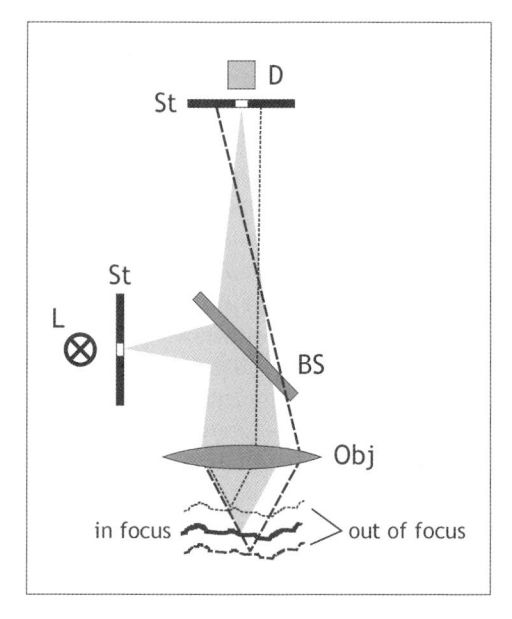

Figure 2.36 Schematic setup of a confocal sensor. L: light source, St: stop, BS: beamsplitter, D: photodetector, Obj.: objective. Dotted and dashed lines: light beams reflected from out-of-focus surfaces.

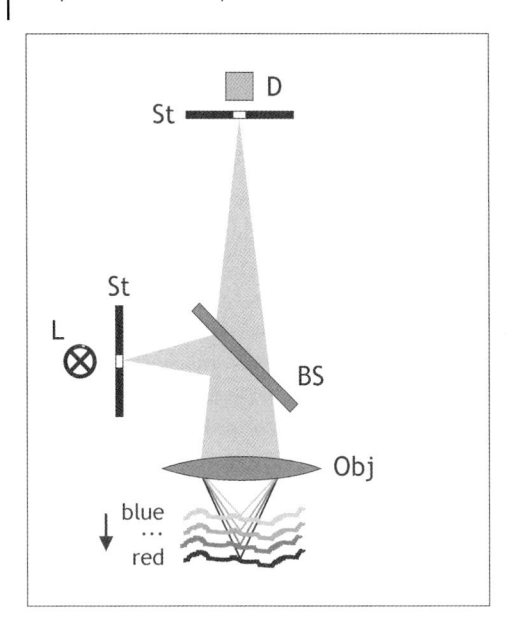

Figure 2.37 Schematic setup of a chromatic confocal sensor. L: light source (halogen lamp), St: stop, BS: beamsplitter, D: photodetector with spectrometer, Obj.: objective. Lines with different grey values indicate different focal lengths for the different wavelengths.

conjugated planes. As a result of this setup, only an object point exactly in focus will yield an image point. Object points that are out of focus will lead to blurred image points and, thus, to low intensities on the detector.

As a consequence, when the sensor performs a vertical (z-) scan at one fixed position, its signal will reach a maximum when the surface point at this position is in focus. Now, the sensor is moved to an adjacent surface point, and a new z-scan is performed. For each of these scans, the focus position is stored, and the complete set of results at all recorded positions gives a three-dimensional representation of the surface.

The drawback of this technology is the necessity to scan the sensor in all three dimensions. The z-scan, however, can be avoided by exploiting a property that is usually regarded as a design error of an optical system: its chromatic aberration (see Section 4.2). In a so-called chromatic confocal setup with a halogen light source and an objective of defined chromatic aberration, there will be a line of vertically staggered foci (Figure 2.37). The focal length, the distance between sensor and focus, is proportional to the wavelength. Thus, the z position of a surface point is coded in the color of the reflected light, which is therefore evaluated with a spectrometer. As in the case of FBGs, this wavelength-measuring technique is far more robust than any technique that would rely on an intensity measurement.

One typical chromatic confocal sensor [70] has a working distance of 6.5 mm and a measurement uncertainty of 0.2 μm at a z resolution of as low as 20 nm. The lateral resolution is 2 μm (Table 2.26). In principle, these values increase for sensors of larger working distances.

Table 2.26 Specifications of a chromatic confocal distance sensor [69].

Property	Value
Working distance	6.5 mm
Distance measurement resolution	20 nm
Measurement uncertainty	0.2 µm
Spot diameter	4 µm
Lateral resolution	2 µm
Measurement angle	90° ± 30° (optical axis to surface)
Max. thickness measurement range	0.9 mm

While the spectrum of light reflected from a solid surface shows one single peak, light from transparent objects, for example, container glasses, will show two peaks that correspond to the two surfaces of the container walls [71]. By evaluating their wavelength difference, it is possible to calculate the wall thickness of the container at the measurement position. The cited reference describes the measurement of wall thicknesses of glass bottles in the production process, and by evaluating only one of the two color peaks, it is also possible to monitor the roundness of the bottles with the same sensor. In the optical industry, it is also possible to measure the center thickness of lenses to monitor their optical specifications during the production process.

6.6.2
Conoscopic Holography

In conoscopic holography [72], the surface is illuminated with the coherent light of a laser spot. The backscattered light passes a lens and a circular polarizer before it enters a uniaxially birefringent crystal (Figure 2.38). The ordinary and extraordinary components of the beam then pass a second circular polarizer. As both components cover different path lengths in the crystal, they interfere at the output

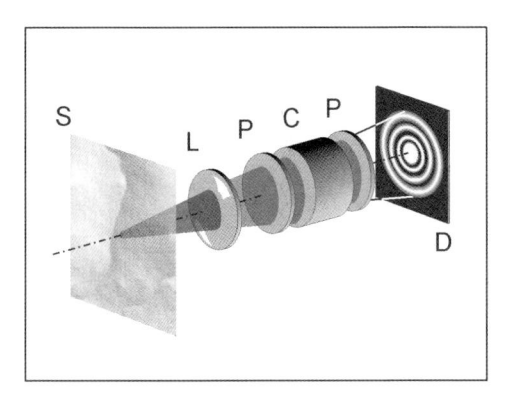

Figure 2.38 Conoscopic holography, schematic setup. S: surface, L: lens, P: polarizer, C: birefringent crystal, D: detector with fringe pattern.

Table 2.27 Specifications of a conoscopic sensor [72].

Property	Value
Measurement range	1.8 mm
Reproducibility	<0.4 µm
Working distance	15 mm
Measuring rate	1 kHz
Measurement angle	90° ± 70° (optical axis to surface)
Lateral resolution	12 µm

of the polarizer, and the output of the optical system is a pattern of concentric interference fringes.

The distance between sensor and surface, d, can be calculated from the radius R_m of the mth interference fringe [72]:

$$d = \sqrt{\left(m \cdot \frac{\lambda}{2}\right)^2 - R_m^2} \qquad (2.40)$$

The great advantage of this optical system is the complete absence of moving parts. Moreover, the system will be insensitive to vibrations due to its collinear design, and it can be adapted to many measurement tasks simply by changing the imaging optics. As the laser beam can be integrated into the optical system, a coaxial illumination can easily be realized. Thus, no shading effects will occur and the setup will also be suitable to scan steep edges of the surface profile.

A typical conoscopic sensor is described in Ref. [73]. Its working distance is 15 mm and it reaches a reproducibility of better than 0.4 µm and a lateral resolution of 12 µm (Table 2.27).

6.6.3
Multiwavelength Interferometry (MWLI)

With no additional parameter input, interferometric distance measurement techniques that use only one wavelength yield unambiguous results only for displacements smaller than one-half of the wavelength. One way to circumvent this problem is to employ more than one wavelength. Suppose we had two wavelengths, λ_1 and λ_2, and that L is the distance between the interferometer and the object. Then, the corresponding phase differences, $\Delta\Phi_1$ and $\Delta\Phi_2$, are (n is the refractive index):

$$\Delta\Phi_1 = \frac{2\pi n}{\lambda_1} \cdot 2L; \; \Delta\Phi_2 = \frac{2\pi n}{\lambda_2} \cdot 2L \qquad (2.41)$$

The "synthetic" phase difference $\Delta\Phi_{12}$ between these two is thus

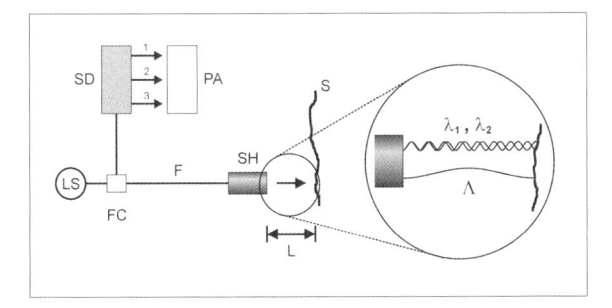

Figure 2.39 Multiwavelength interferometry (MWLI), schematic setup. LS: light source, FC: fiber coupler, F: fiber, SH: sensor head, L: distance to surface, S: surface, SD: spectral filtering and distribution, PA: phase detection and analysis. λ_1, λ_2: detection wavelengths, Λ: synthetic wavelength.

$$\Delta\Phi_{12} = \Delta\Phi_1 - \Delta\Phi_2 = 2\pi n \left(\frac{1}{\lambda_1} - \frac{1}{\lambda_2} \right) \cdot 2L = \frac{2\pi n}{\Lambda} \cdot 2L \qquad (2.42)$$

As a consequence, $\Delta\Phi_{12}$ depends on a new wavelength, the synthetic wavelength Λ, which is given as

$$\Lambda = \frac{\lambda_1 \lambda_2}{\lambda_1 - \lambda_2} \qquad (2.43)$$

If the difference between the two wavelengths is small, Λ will be comparably large, and this increases the unambiguity range of the interferometer (Figure 2.39). In the design of a newly developed instrument [74], three wavelengths are coupled into a fiber probe with which the surface is illuminated. The reflected light is picked up by the same fiber and the wavelengths are separated by filter systems. Thus, the distance to the object can be determined with a precision of about $\lambda/2000$ across an unambiguous working range of one-half of the wavelength when the phase of only one of the wavelengths is evaluated. For a wavelength of 1550 nm, the precision is thus in the nanometer range. In addition, the instrument evaluates the synthetic phase and is therefore capable of maintaining its precision over a total working range of as large as 2 mm.

The above-mentioned MWLI sensor works at a wavelength of (1550 ± 40) nm and features a working range of 0.5 mm to 1 m and an absolute measurement range of 2 mm to 20 cm. Within its unambiguousness range of 2 mm, it reaches a longitudinal resolution of between 0.5 and 10 nm. The sampling rate is given as 4 kHz ([75], Table 2.28).

6.6.4
White-Light Interferometry

The basic setup of a white-light interferometer corresponds to that of a Michelson interferometer: A collimated light beam is split up into two partial beams that

Table 2.28 Specifications of a MWLI sensor [74].

Property	Value
Measurement range (absolute)	2 mm–20 cm
Unambiguousness range	2 mm
Working distance	0.5–1 mm
Measuring rate	1–5 kHz
Measurement wavelength	1550 ± 40 nm
Image field diameter	30–50 µm

travel some distance before they are reflected back to the beamsplitter by two mirrors. The intensity of the output beam that leaves the beamsplitter is determined by the path difference between the partial rays in the two interferometer arms. When both are of equal length, the path difference is zero, and the intensity reaches its maximum. If the length of one of the two arms changes continuously, the intensity will vary periodically, and its period will correspond to one-half of the wavelength of the light source [3]. The displacement range over which the interferences will be visible defines the coherence length of the light source, see Eq. (1.8). A particular setup is the white-light interferometer, or coherence radar [76]. It differs from the original setup only in that the light source is a broadband light source with a small coherence length. As the white-light interferometer is employed for the measurement of surface profiles, one of the two mirrors is replaced by the surface to be investigated (Figure 2.40).

As the bandwidth of the light source is large, its coherence length is small, and interferences become visible only at about equal lengths of the two interferometer arms. The intensity variation, I, of the central spot of the interference pattern as a function of the displacement Δz of one of the two arms (Figure 2.40) is

$$I(\Delta z) = I_0 E(\Delta z)\left(1 + \cos\left(\frac{2\pi \Delta z}{\lambda}\right)\right) \tag{2.44}$$

Here, the width of the envelope, $E(\Delta z)$, is determined by the coherence length of the light source, and the center of the envelope marks the position where both interferometer arms have equal lengths.

The coherence length of a broadband light source, for example, of an LED, can be as small as only 10 µm. Thus, the visibility of the interference fringes indicates the condition that both interferometer arms are of equal length with high sensitivity. The setup in Figure 2.40 is designed to perform such an interferometric measurement over a comparably large area in one shot. The signals are recorded by a CCD camera in two dimensions simultaneously. When the surface under investigation is scanned into the direction of the incident light, that is, when a z-scan is performed, the two-dimensional distribution of the z positions of the interference maxima across the CCD pixels will yield a high-resolution representation of the surface.

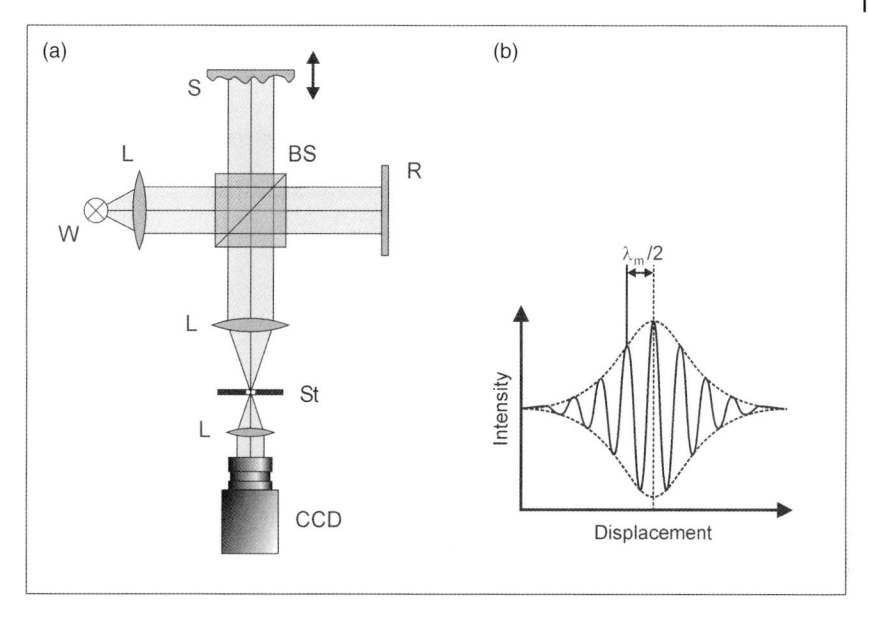

Figure 2.40 White-light interferometer. (a) schematic setup, W: broadband light source, L: lens, BS: beamsplitter cube, R: reference mirror, S: sample surface, St: stop, CCD: camera. (b) Signal form, λ_m: mean wavelength of the light source.

Table 2.29 Specifications of a white-light interferometer [76].

Property	Value
Measurement range	0–500 µm
Resolution	<40 nm
Image field diameter	19 mm
Lateral resolution	9–39 µm
Operating temperature	+5 °C to +35 °C

A typical commercial instrument works with image field diameters of up to 19 mm and the lateral resolution is significantly below 50 µm, depending on the size of the image field (Table 2.29). At a vertical measurement range of 500 µm, the vertical resolution is less than 40 nm [77].

A particular application for white-light interferometry is optical coherence tomography (OCT). This technique makes use of the fact that near-infrared light can reach penetration depths of up to 2 mm in tissue. It is therefore applied, for example, in ophthalmology where cross-sections of the human retina can be recorded with an axial resolution of 5 µm and a transverse resolution of 15 µm [78].

6.6.5
Near-Field Optical Microscopy

The resolving power of a conventional optical microscope is determined by the illumination wavelength and by the numerical apertures of its condenser (NA_{cond}) and objective (NA_{Obj}). The numerical aperture contains both the half angle α of the emitted (condenser) and received (received) light cones, and the refractive index, n, of the medium outside the respective optical element: $NA = n \sin\alpha$. This is described by the well-known Abbe formula for the minimum distance Δx that can be resolved [79]:

$$\Delta x = \frac{\lambda}{NA_{Cond} + NA_{Obj}} \quad (2.45)$$

As an example, a microscope with a tungsten halogen lamp with a central wavelength of about 550 nm and $NA_{cond} = NA_{Obj} = 1.25$ can resolve structural dimensions of about 220 nm. This value can be reduced by increasing the numerical apertures or by decreasing the illumination wavelength. For visible light, however, there is a short-wavelength limit at about 400 nm, and the numerical apertures cannot be increased to values significantly larger than 1.5. This was the direct motivation for the development of the electron microscope by Ernst Ruska and Max Knoll in 1931 with which structures less than 1 nm in size can be observed.

There are, however, optical techniques with increased optical resolution: The scanning near-field optical microscope (SNOM) captures the structure of a surface with an ultrahigh resolution that exceeds the value given in Eq. (2.45). The sample is illuminated through an aperture of a diameter that is significantly smaller than the illumination wavelength. This was first demonstrated for visible light in the "optical stethoscope" [80] with a resolution of about 50 nm, which is only about one-tenth of the wavelength of visible light.

Figure 2.41 shows the principle: a single-mode optical fiber is drawn to a tip with a diameter of less than 100 nm and coated with aluminum. Thus, a subwavelength aperture forms at the apex of the fiber tip. The fiber is brought into a distance of between 1 and 100 nm from the surface. There are two modes of operation: the fiber can either be used in the transmission mode to illuminate an area of subwavelength size, or in the collection mode to receive light from an area of subwavelength size.

Outside the fiber aperture, there is only an evanescent light field due to the small aperture size. This evanescent field leads to a light concentration onto a spot of about the aperture size. The near-field, whose intensity decreases exponentially with the distance from the aperture, allows to measure the surface topography with highest resolution and below the diffraction limit of the detection optics (symbol D in Figure 2.41). The fiber is scanned with a three-dimensional piezotransducer. The resolution of the surface image is limited only by high spatial frequencies of the intensity distribution in the aperture. A commercial instrument

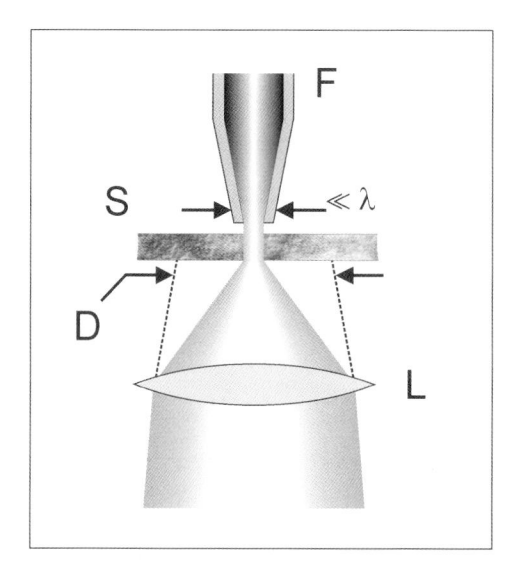

Figure 2.41 Scanning near-field optical microscope (SNOM). F: aluminum-coated fiber tip, S: sample surface, L: far-field lens, D: lens diffraction limit.

[81] covers a scan range of 100 μm × 100 μm × 20 μm with an optical resolution of 100 nm, depending on the aperture size.

6.6.6
Contouring: Structured-Light Techniques

A very straightforward way to assess the contour of a surface or an object is to project a deterministic, known structure, for example, a laser line, onto it. The line is distorted by the surface profile, with the amount of distortion also depending on the angle between line projector and viewing direction. Today, smart cameras have become available not only to capture the image of the projected line but also to calculate the profile on-board [82]. To capture the surface profile in two dimensions, the surface would have to be scanned into the direction perpendicular to the line. There is, however, also the possibility to project a grid of lines – or another two-dimensional structure of known geometry. One very modern embodiment of such a projector employs micromirror devices. They allow very fast scans of an object and are, for example, employed in scanners for parts of the human body [83]: the scanner captures 80 000 object points during one scan interval of only 0.025 s. Moreover, the accuracy of the captured profile is better than 0.3 mm (Table 2.30).

Moiré contouring techniques are one particular layout for projection techniques. They have the great advantage that they capture the surface profile over a large field in one shot, but at the cost of limited resolution. The setup for a shadow Moiré is quite simple (Figure 2.42): a contoured surface is illuminated through a

6 Optical Sensor Concepts

Table 2.30 Specifications of a 3D body scanner [82].

Property	Value
Number of xyz coordinates per scan	80 000
Scan time	0.025 s
Angular measurement range	360°
Measurement uncertainty	<0.3 mm

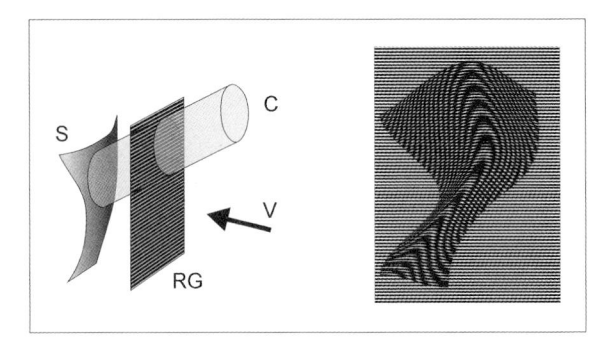

Figure 2.42 Moiré contouring. S: contoured surface, C: collimated light, RG: Ronchi grid, V: viewing direction.

grating or line ruling, a so-called Ronchi grid. The ideal light source for this application is a point source with collimating optics. When the surface is observed through the grid, the received image is composed of both the grating and its shadow on the surface. The result of this addition of grid and shadow patterns is a Moiré contour map (Figure 2.42, [84]). Let us assume that each of the two grids has a sinusoidal intensity profile I_1, I_2, and that the two grids are parallel (k_1, k_2 are angular wavenumbers and x the propagation direction):

$$I_1(x) = I_0 \sin(k_1 x), \quad I_2(x) = I_0 \sin(k_2 x) \tag{2.46}$$

After some mathematical manipulation, the resulting pattern, I_{res}, is of the form

$$I_{res}(x) = I_1(x) + I_2(x) = 2I_0 \cos\left(\frac{k_1 - k_2}{2} x\right) \cdot \sin\left(\frac{k_1 + k_2}{2} x\right) \tag{2.47}$$

This mathematical operation is very similar to the one that is applied to the interference of waves in interferometry. Moiré patterns are also the basis for the principle of operation of incremental encoders (Figure 2.9). The two interfering patterns may also be tilted against each other, and the resulting pattern shows lines into the direction of the bisecting line of the angle between the two. The resulting beat frequencies determine the low-frequency components in the Moiré contour plot.

Moiré contouring is inexpensive, easy to apply, and yields quantitative contouring data. One interesting application for this technique is the analysis and study of artifacts in archeometry [85]. Here, contours, shapes, and possible irregularities can easily be determined by calibrating the setup with an object of known size and shape. A commercial contouring instrument [86] can capture an area of 600 mm × 600 mm with a resolution of 2.5 μm.

A wide variety of other techniques exists in the field of contouring and assessment of three-dimensional object data that all make use of structured-light techniques. Besides the above-mentioned micromirror devices, there are also special coding techniques, for example, with Gray-coded light patterns. These techniques are, however, very sophisticated and shall therefore not be discussed in further detail in the framework of this textbook.

6.6.7
Concepts: Cross-Correlation Analysis and 2D Fourier-Transform Techniques

The high-resolution techniques for the analysis of technical surfaces described so far have comparably elaborate setups. All techniques that utilize image-processing systems may also require considerable computational power, particularly in the case of high-speed applications. One example is the inspection of strip materials during the production process, and to date, a number of ideas exists to replace the electronic evaluation of the optical signals with a purely optical one. This would not only lead to a signal evaluation virtually at the speed of light, but also to significantly reduced specifications for the electronic circuitry. To the author's knowledge, none of these concepts has become a commercial product so far, but they shall be mentioned here to show what is or may be possible with optical sensors in the (near) future.

One approach is based on the cross-correlation sensors described in Section 6.2.3. There, Eq. (2.10) describes their output signal as the cross-correlation between the two-dimensional intensity distribution of the surface image, $i(x, y)$, and the aperture function (of the grating), $a(x, y)$. With the relative velocity v between sensor and surface, the sensor output $S(t)$ is

$$S(t) = \iint_A a(x, y) \cdot i(x - vt, y) \, dx \, dy \tag{2.48}$$

The spectral power density of this signal yields its frequency spectrum, and it follows from the convolution theorem that this frequency spectrum equals the product of the spectral power densities $I(x, y)$ of the image and $A(x, y)$ of the aperture function:

$$C(f/v) = \text{const.} \cdot A(f/v) \cdot I(f/v) = T(f/v) \cdot I(f/v) \tag{2.49}$$

Here, the argument f/v has the dimension of an inverse length so that $A(f/v)$ and $I(f/v)$ describe the spatial frequency spectra of surface and aperture, respectively. Apart from a constant factor, A can be understood as the transfer function of the sensor. This transfer function, which is a constant for every individual sensor, can be measured by recording the correlation signal of a known surface structure [21] so that the spatial frequency spectrum of an unknown surface can simply be calculated by dividing the power spectrum of the sensor signal by the transfer function. As the spatial frequency is just a different representation of the surface profile, averaged over the sensor's field of view, it may contain characteristic surface parameters for an analysis of the surface.

A similar technique utilizes the direct Fourier transform of a surface image. It is the basic principle of imaging optics that the intensity distribution in the back focal plane of an objective equals the two-dimensional Fourier transform of the aperture function, multiplied with the intensity distribution of the image (see, e.g., [87]). If $U_i(x, y)$ is the intensity distribution in the image plane, $P(x, y)$, the aperture function (= 1 inside the lens aperture, = 0 otherwise) and f the focal length of the lens, the complex intensity distribution in the back focal plane, $U_f(x_f, y_f)$, can be written as [88]

$$U_f(x_f, y_f) = \frac{e^{i\frac{k}{2f}(x_f^2 + y_f^2)}}{i\lambda f} \iint_{-\infty}^{\infty} U_i(x, y) P(x, y) e^{-i\frac{2\pi}{\lambda f}(x x_f + y y_f)} \, dx \, dy \qquad (2.50)$$

Thus, an objective can be understood as an optical computer that calculates the Fourier transform of the product of aperture function and image intensity distribution at the speed of light. Changes in the surface properties will result in changes in the Fourier spectrum, that is, in the spatial frequency spectrum of the image.

The authors of Ref. [88] applied this technique for the surface analysis of paper during its production process. The paper surface usually has a homogeneous but stochastic structure, and this corresponds to a spatial frequency spectrum that follows a $1/f^\alpha$ characteristic. Figure 2.43 shows schematic representations of two 2D spatial frequency spectra in order to illustrate the sensor principle.

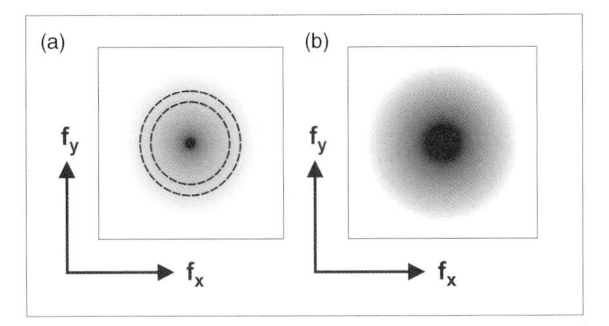

Figure 2.43 Schematic illustration of the 2D spatial frequency spectrum of paper surface images, optically computed in the back focal plane of an imaging lens. Amplitudes of the frequency components represented by grayscale values. (a) Good paper formation with a ring mask (dashed lines) for the analysis of spatial frequency intervals, (b) bad paper formation.

Table 2.31 Specifications of a 2D Fourier processor prototype for surface analyses [87].

Property	Value
Wavelength of light source	670 nm
Surface velocity	600–2000 m/min
Maximum image size	832 pixels × 632 pixels
Maximum measurement frequency	200 Hz
Dynamic range	12 bits

One important parameter in the production of paper is its formation: the finer the paper structure, the better its formation. This property is represented in the spatial frequency spectrum in a very compact manner; in Figure 2.43, the amplitudes of the spatial frequency components are represented by a grayscale value. In case of a good paper formation, there are only small amplitudes at small spatial frequencies. If the formation deteriorates, the frequency distribution becomes broader. In order to analyze these images, amplitude information can be read out in ring-shaped regions of interest.

As this technique reduces the amount of data that has to be recorded, and also the amount of data that has to be evaluated to a minimum, surface analyses may be carried out at surface velocities of up to 2000 m/min. Also, as the largest and most decisive part of the computing is done in the optical system of the sensor, the electronic circuit can be realized with a little more than one digital signal processor (DSP). Table 2.31 shows the specifications of a sensor prototype.

6.7
Deformation and Vibration Analysis

Vibration analysis is of high importance in many applications, for example, in the automotive industry where these analyses contribute to the optimization of stability and riding comfort. Whenever it is sufficient to measure vibrations in only one dimension, the measurement task can be reduced to a highly dynamic determination and frequency analysis of the surface velocity into the direction that is of interest. If, however, three-dimensional vibrations must be analyzed, or if the vibration behavior of an object is not known, either a vector-based technique with at least three one-dimensional sensors or a whole-field technique is applied.

6.7.1
Laser Vibrometers

Laser Doppler vibrometers utilize the Doppler effect for the determination of object movements into the direction of a laser beam, very similar to the modulator described in Section 4.7. The Doppler effect describes the influence of the

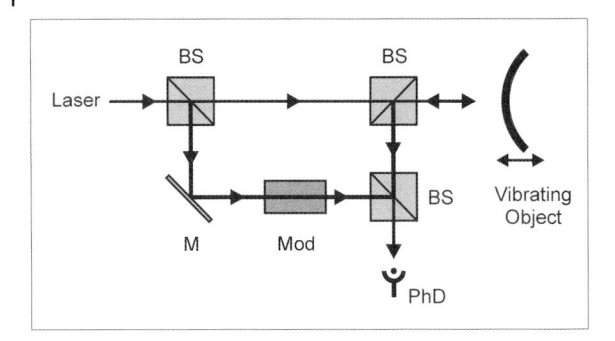

Figure 2.44 Mach–Zehnder interferometer for laser vibrometry. BS: beamsplitter cube, M: mirror, Mod: modulator, PhD: photodetector.

movement of an object with velocity v on the frequency f_0 of the reflected light. Under the condition that $v \ll c$, the resulting frequency f' can be calculated to

$$f' = f_0 \sqrt{\frac{c \pm v}{c \mp v}} \qquad (2.51)$$

The upper signs in numerator and denominator describe the case when light source and observer approach, the lower signs hold when their distance increases. When the object vibrates, the light frequency will change periodically with the object movement. This will lead to a periodical phase shift between incident and reflected beams that can favorably be evaluated by an interferometer.

Figure 2.44 shows the schematic setup of a Mach–Zehnder-type interferometer. The incoming light from a laser, typically a HeNe laser with an emission wavelength of 632.8 nm, is split up by a beamsplitter cube. One partial beam is directed to the object under investigation, the other passes a reference arm and is eventually folded with the beam that is reflected by the object.

When the object moves with respect to the vibrometer, the phase shift between reference and measurement beam leads to a change in the interference pattern on the photodetector. When the interference pattern shifts by exactly one interference fringe, this corresponds to a path length difference of one wavelength. Sophisticated interpolation of the interference fringes will increase the resolution even further so that resolutions in the nanometer range become possible.

The absolute value of the object displacement is determined by counting the number of interference fringes. The sign of this displacement is determined by adding a modulator to the reference arm of the interferometer, usually a Bragg cell (see Section 4.7) that shifts the light frequency by some 10 MHz. In case the object does not move, the detector will receive exactly the modulation frequency. When the object vibrates, its movement will lead to an increase or decrease of the modulation frequency, depending on the direction of the movement.

A typical commercial laser vibrometer has a bandwidth of up to 80 kHz and it can measure velocities of up to 20 m/s. The sample sizes may range from some

Table 2.32 Specifications of a laser vibrometer [88].

Property	Value
Measurement range	>0.4–100 m
Sample size	Mm² to m² range
Number of channels	Up to 8
Scanning frequency	80 kHz
Maximum vibration velocity	20 m/s

square millimeters to some square meter while the standoff distances can be as large as 100 m (Table 2.32, [89]). When three of this vibrometers are combined to a 3D analysis system, they can perform a full-field, 3D vibration analysis as the result of a geometrical scan of the object over a grid of 512 × 512 points.

6.7.2
Speckle-Pattern Interferometry

Whenever laser light is reflected from an optically rough surface, the high degree of temporal and spatial coherence will result in a random but also stationary intensity distribution. The intensity in each point of this so-called speckle pattern is determined by the phase differences of the single waves that interfere in this point. The effect will be most visible for a surface roughness that is of the order of the light wavelength. Once again, and for the example of two waves with intensities I_1 and I_2 that interfere with a phase difference of $\Delta\Phi$, the resulting intensity I_{12} is [3]

$$I_{12} = I_1 + I_2 + 2\sqrt{I_1 I_2}\cos\Delta\Phi \qquad (2.52)$$

If both waves are of equal intensities ($I_2 = I_1$), this reduces to [90]:

$$I_{12} = 2I_1(1+\cos\Delta\Phi) = 4I_1 \cos^2\left(\frac{\Delta\Phi}{2}\right) \qquad (2.53)$$

When the phase difference between the two waves equals zero or an even integer of π, the resulting intensity reaches its maximum of $I_{12,\max} = 4\,I_1$. When it equals an odd integer of π, the speckle pattern will show a dark spot because I_{12} becomes zero.

The speckle pattern will also contain information about the structure and contour of the surface. However, simply recording a second speckle pattern after some (short) time interval and subtracting it from the first pattern will not yield any information about the changes of the surface during vibration. For out-of-plane vibrations into the direction of the camera's line of sight, the solution is to extract a reference beam from the same laser and to fold it with the object beam to form an interference pattern on a CCD camera (Figure 2.45). The result is also a speckle pattern on the camera chip. This so-called holographic

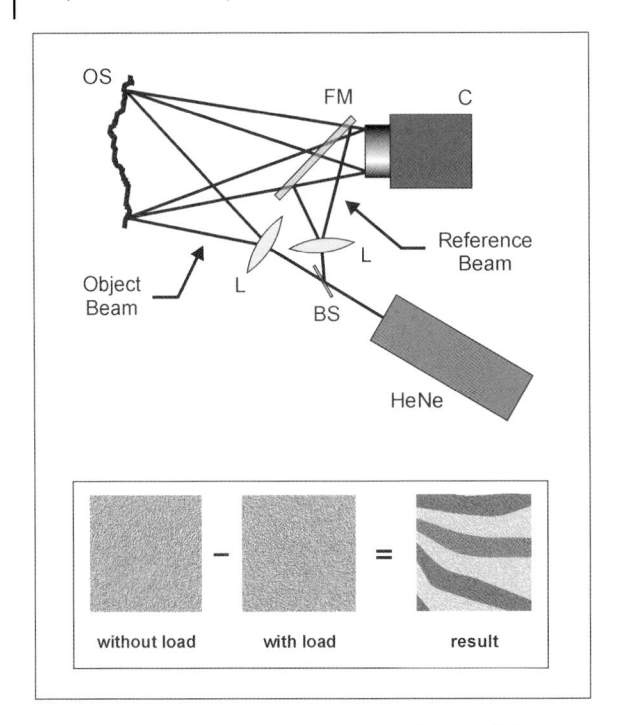

Figure 2.45 Speckle-pattern interferometry for out-of-plane vibrations, setup and resulting fringe pattern (schematic). OS: object surface, L: lens, BS: beamsplitter, HeNe: HeNe laser, FM: folding mirror, C: CCD camera.

speckle pattern does not only contain the amplitudes, but also the phases of each speckle.

When the surface changes, so will the phases of the light waves in the object beam and a different speckle pattern results. The evaluation then calculates the point-to-point differences of two subsequent images: as a result, all image points for which the phase differences are multiples of 2π do not change their intensity values, and their differences become zero. On the other hand, all image points for which the phase differences are odd multiples of π will change from bright to dark (or vice versa) and lead to maximum intensities in the difference image. The result is a difference image that shows bright and dark contour lines illustrating the way that the surface distorts during vibration. As the difference between a bright and a dark line corresponds to a path length difference of $\lambda/2$, this technique has a resolution of one-half of the employed wavelength. For a HeNe laser, the resolution will thus be about 316 nm.

As this technique relies on the electronic evaluation of the speckle patterns, it is denoted by "electronic speckle pattern interferometry" (ESPI). The experimental setup of this full-field interferometric technique closely resembles that of white-light interferometry. The difference is that the fringes can only be depicted when the two speckle patterns are subtracted from each other. Thus, the fringes are no

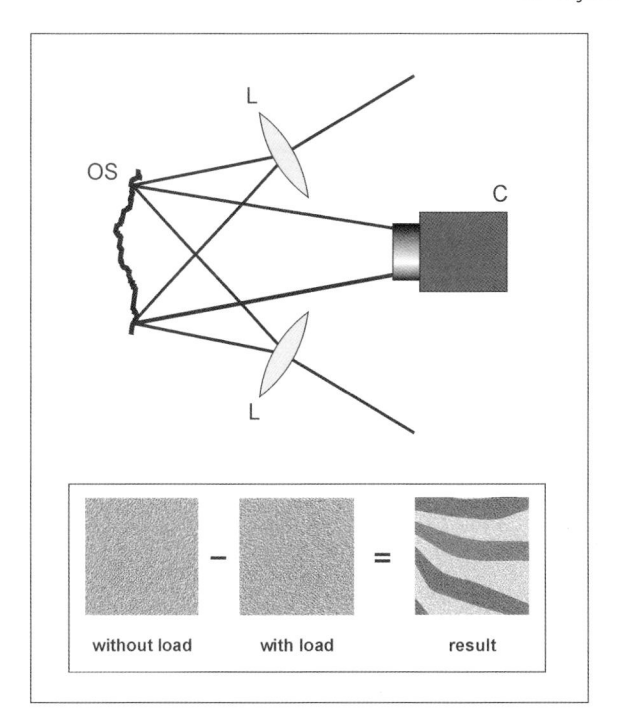

Figure 2.46 Speckle-pattern interferometry for in-plane vibrations, setup and resulting fringe pattern (schematic). OS: object surface, L: lens, C: CCD camera.

interference fringes in the original sense – they are usually referred to as "correlation" fringes. As two fringes represent an absolute phase difference of a multiple of 2π between object and reference beam, the sign of the displacement is not unambiguous. This ambiguity can be resolved, for example, by introducing defined phase shifts between the two beams with a modulator.

As the setup for out-of-plane ESPI measurements is basically a setup for a full-field displacement measurement, it is unsensitive against displacements and vibrations perpendicular to the camera's line of sight. Figure 2.46 shows a schematic setup for the ESPI measurement of these in-plane variations: the surface is illuminated by two expanded beams from the same laser source. The beams are incident from different directions, but under the same angles to the surface normal. Here, the speckle hologram is formed on the surface itself and as in the case of out-of-plane measurements, the subtraction of two subsequently recorded images will result in a fringe pattern that depicts the contours of the deformed object.

A commercial instrument [91] analyzes the vibration-induced deformations of a surface over an area of 200 mm × 300 mm with a displacement accuracy of up to 0.03 μm. Its working distance ranges from 0.2 to 1.0 m (Table 2.33).

A special embodiment of ESPI is realized by extracting the reference beam from the object beam itself. This is usually done by imaging the surface twice onto the

Table 2.33 Specifications of an ESPI vibrometer [90].

Property	Value
Measurement area	200 mm × 300 mm
Working distance	0.2–1 m
Displacement measurement uncertainty	0.03–0.1 μm
Data acquisition speed	3.5 s (one measurement step, 3D analysis)
Measurement wavelength	785 nm

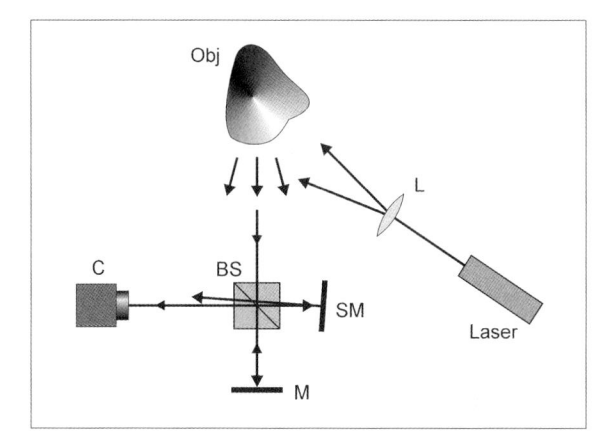

Figure 2.47 Shearography, schematic setup. L: lens, Obj: object, BS: beamsplitter, SM: shearing mirror, M: mirror, C: CCD camera.

sensor, but the second image with a slight lateral shift. This so-called shear is optically introduced with a shearing element, for example, a wedge in front of the camera lens [92] or with a slightly tilted mirror in a Michelson-type interferometer (Figure 2.47). With an untilted mirror, this setup could also be used for out-of-plane ESPI measurements.

While the object experiences deformations, two speckle patterns are recorded, and as in ESPI, the difference between these two patterns yields information about the deformation. The inherent self-referencing of this technique, which is also denoted by electronic speckle pattern shearing interferometry (ESPSI), allows a compact setup and is therefore commercialized by several manufacturers [93].

6.7.3
Holographic Interferometry

While the superposition of object and reference beam in ESPI leads to an interference image only after a digital processing of two subsequently recorded images,

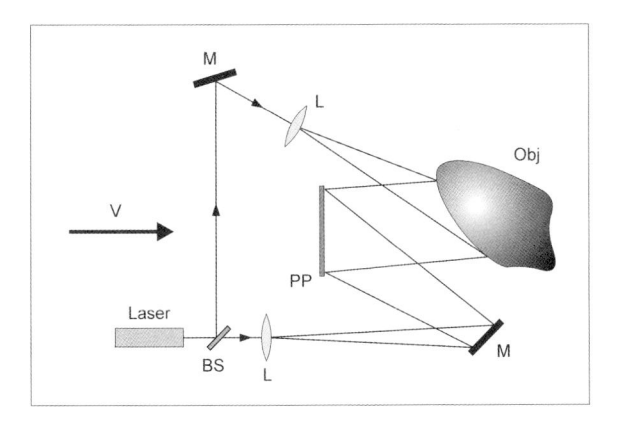

Figure 2.48 Holographic interferometry, schematic setup. BS: beamsplitter, L: lens, M: mirror, PP: photo plate, Obj: object, V: viewing direction when reconstructing the hologram with the reference beam alone.

holographic interferometry directly yields a contour image as a depiction of a vibration- or load-induced deformation of an object. The basic setup is the one that is usually employed for holography (Figure 2.48): the object under investigation is illuminated by a diverging laser beam, the object beam. It is mathematically represented by a wave, ψ_{obj}, with amplitude $\psi_{obj,0}$, wave vector k, frequency ω and phase Φ_{Obj}:

$$\psi_{obj}(r,t) = \psi_{obj,0} \cdot e^{i(kr - \omega t + \varphi_{obj})} \tag{2.54}$$

The undisturbed, but also expanded reference beam is accordingly represented by

$$\psi_{ref}(r,t) = \psi_{ref,0} \cdot e^{i(kr - \omega t + \varphi_{ref})} \tag{2.55}$$

Both beams are brought into superposition on a photo plate. When the photo plate is developed, it contains a speckle pattern as the result of the interference between the two beams. From Eqs. (2.54) and (2.55), the intensity distribution $I(x,y)$ on the photo plate is the square of the absolute value of the sum of the two wavefunctions:

$$I(x,y) = \psi_{obj,0}^2 + \psi_{ref,0}^2 + 2\psi_{obj,0}\psi_{ref,0} \cdot \cos(\varphi_{obj} - \varphi_{ref}) \tag{2.56}$$

This expression is the same as the one already given in Eq. (2.52). This so-called hologram thus does not only contain information about the amplitudes of the backscattered light, but also about the relative phases between object and reference beam, that is, about the shape of the object.

The reconstruction process requires a setup similar to that of Figure 2.48, but without object and object beam: only the reference beam is required and now acts as the reconstruction beam, ψ_{rec}:

$$\psi_{rec}(r,t) = \psi_{rec,0} \cdot e^{i(kr - \omega t)} \tag{2.57}$$

The reconstruction wave propagates through the hologram, and the resulting wave of this reconstruction, ψ_{result}, is proportional to the product of the reconstruction wave with the hologram's intensity distribution $I(x,y)$. The calculation yields:

$$\psi_{\text{result}} = \psi_{\text{result},1} + \psi_{\text{result},2} + \psi_{\text{result},3} \tag{2.58}$$

with

$$\psi_{\text{result},1} \propto \left[\psi_{\text{obj},0}^2 + \psi_{\text{ref},0}^2\right]\psi_{\text{ref},0} \cdot e^{i(\varphi_{\text{ref}} - \omega t)} \tag{2.59}$$

$$\psi_{\text{result},2} \propto \psi_{\text{ref},0}^2 \cdot \psi_{\text{obj},0} \cdot e^{i(2\varphi_{\text{ref}} - \varphi_{\text{obj}} - \omega t)} \tag{2.60}$$

$$\psi_{\text{result},3} \propto \psi_{\text{ref},0}^2 \cdot \psi_{\text{obj},0} \cdot e^{i(\varphi_{\text{obj}} - \omega t)} \tag{2.61}$$

The result of the reconstruction is composed of three components. The first component, $\psi_{\text{result},1}$, is simply the amplitude-modulated reference wave. It can be understood as the zero-order interference term of the hologram grating and contains no phase information of the object. $\psi_{\text{result},2}$ and $\psi_{\text{result},3}$, on the other hand, are first-order interference terms with amplitudes proportional to that of the object wave. The first of these two terms, however, contains the negative value of its phase. This component therefore reconstructs the object, but in a somewhat inverted manner: When it is viewed, it seems to be three-dimensional, but the depth offset of the image points is inverted, and the observer does not see a true three-dimensional representation of the object. The last term closely resembles the object wave, and this time with the correct sign of its phase. When viewing this wave, the observer receives a phase distribution that corresponds to that of the object, as if it were at the position where it was recorded. The hologram is thus an optical storage for the three-dimensional shape of an object.

When the object changes its form, for example, due to external forces, its hologram will change as well. To evaluate this deformation, this hologram can be made to interfere with the original one by placing it in the position of the photo plate. In its plane, the two interference patterns combine and when this superposition is viewed in the same way that the hologram is viewed when it is reconstructed, the result is a holographic image with an additional interference pattern that shows the contour map of the deformed object. As in all interferometric images, two fringes correspond to a difference in object distance of one wavelength, so that this technique is quite sensitive. There are basically three different techniques to record the interferometric image in holographic interferometry:

- In the double-exposure technique, the photo plate is exposed twice, that is, with the two different holograms. When the image is reconstructed, it contains both the three-dimensional image of the object and the contour map with the deformation information.

- With the real-time technique, the photo plate is exposed only once, and when the object experiences deformations, they can be observed visually.

- For vibration analyses, a time-averaged hologram is recorded with an exposure time that is long compared to the oscillation period: wherever the object has oscillation nodes, it is in rest relative to the photo plate – these areas of the hologram will show bright spots. At the positions of the oscillation amplitudes, however, the phase between object and reference beam is in constant change, and the hologram will show dark areas.

In contrast to ESP(S)I, the setup for holographic interferometry has the disadvantage that it cannot be realized in one compact instrument. Although this technique is very sensitive and results in a direct image of an object's deformation, there is no "holographic interferometer" available by the time that this manuscript was finished – at least to the author's knowledge. For more information about holographic interferometry and its application, the interested reader is referred to Ref. [94].

6.8
Wavefront Sensing and Adaptive Optics

When a light wave propagates through a medium, a wavefront is the set of all points of the wave that are of identical phase, or in other words that have traveled the same time of flight from the light source. The wave vectors of all points are normal to the wavefront and define the direction of the "light rays" in the model of geometrical optics.

Just two examples to illustrate this concept: perfectly collimated light can be described as a plane wave or as a set of parallel beams. Thus, the wavefronts are parallel planes perpendicular to the direction of propagation. Light from a point source, for example, from a distant star in astronomy, has spherical wavefronts (Figure 2.49). When light propagates through a homogeneous and isotropic medium, the wavefronts remain undistorted, but in real life, this is not always the case. A good example from everyday experience is the flickering image of objects that are observed through a hot air volume. It can, for example, be observed right above hot road surfaces or motorcar roofs. Astronomers know the effect as "seeing," and it occurs, for example, when the line of sight to an object passes a rooftop closely. The temperature difference between the cool air and the comparably warm rooftop leads to fluctuations in the air density and, thus, in the refractive index which will then lead to the wavefront distortions and, thus, to the well-known flickering of stars in the night sky. Just one remark: although a distant star is observed as a point-like object, the arriving wave can be considered a plane wave.

There are several ways to avoid wavefront distortions: The easiest workaround is to only observe objects with high elevations above the horizon, which is not always acceptable or possible. Another solution is the installation of telescopes in the Earth's orbit. The only space telescope that has been launched so far is the Hubble Space Telescope, operated by NASA. It has been in service since 1990 and circles Earth in an altitude of about 590 km above ground. Although the absence of an atmosphere excludes atmospheric seeing, this solution is obviously very expensive and complex.

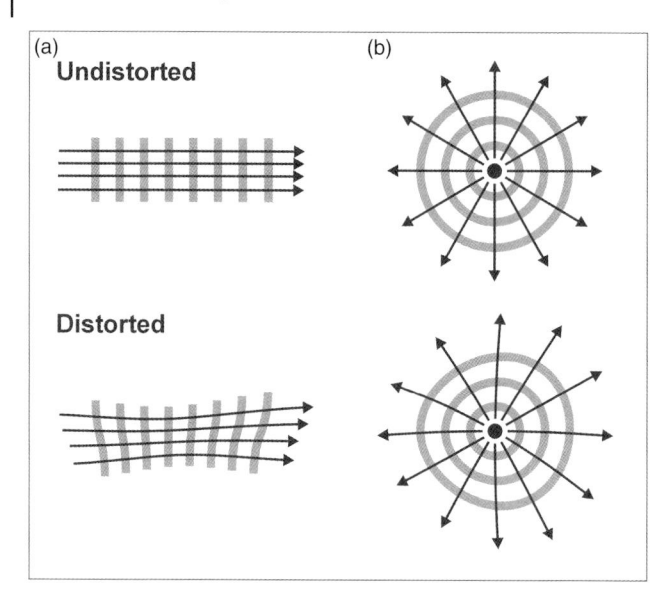

Figure 2.49 Undisturbed and distorted wavefronts for a plane wave (a) and for a spherical wave (b).

An alternative that can be employed on Earth makes use of a sensor that captures the shape of the received wavefront and drives an adaptive system that compensates the wavefront distortions. A widely used sensor for this purpose is the Shack–Hartmann sensor (Figure 2.50). It basically consists of an imaging detector (CMOS or CCD) with a lenslet array. The distance between detector and array corresponds to the focal distance of the lenslets. When a plane wave arrives at the sensor, every lenslet forms an image point on the detector. As the wave vectors are all parallel in the ideal case, the image points are equidistant on a two-dimensional grid.

Any distortion of the wavefronts will lead to a distortion of the image and, thus, to a shift in the position of the image points. When the ideal positions of these image points are known, for example, by observing a bright natural guide star (NGS) or by sending a laser beam into the direction of observation in order to produce an artificial star (laser guide star, LGS), the wavefront distortion can be evaluated with a mathematical formalism based on Zernike polynomials. Shack–Hartmann sensors are available as integrated components. One of these [95] works in a wavelength range between 193 nm and 1064 nm and reaches a measurement uncertainty between $\lambda/50$ and $\lambda/250$ (Table 2.34).

For the correction of the wavefront distortions, it is important to find a theoretical model of the impact of atmospheric turbulences on light propagation. In the frame of the Kolmogorov model, a so-called Fried parameter r_0 is defined that describes the maximum aperture diameter, up to which an optical telescope works only diffraction-limited. It is (see [96] and references cited therein):

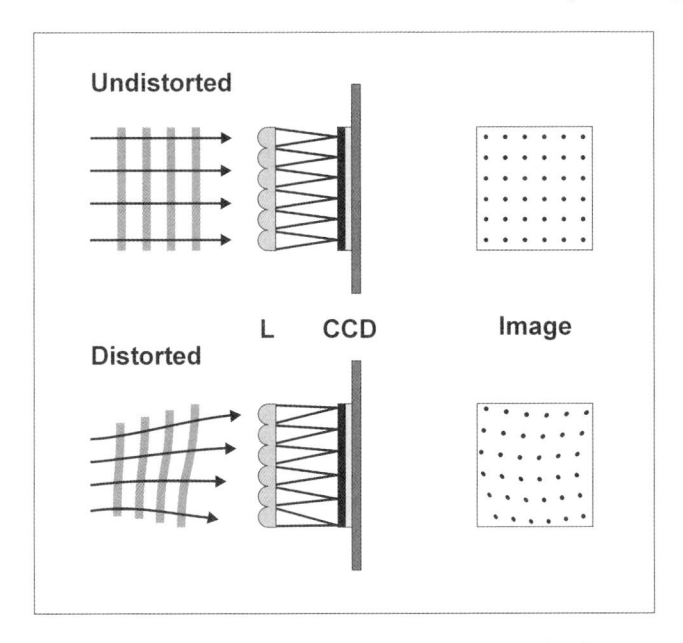

Figure 2.50 The Shack–Hartmann sensor (schematic). L: lenslet array, CCD: imaging detector.

Table 2.34 Specifications of a Shack–Hartmann wavefront sensor [94].

Property	Value
Wavelength range	193–1064 nm
Measurement uncertainty	$\lambda/50$–$\lambda/250$
Dynamic range	λ–100 λ (at 632.8 nm wavelength)

$$r_0 = \left(0.423 \left(\frac{2\pi}{\lambda} \right)^2 \frac{1}{\cos z} \int_0^{H_{max}} C_n^2(h)\,dh \right)^{-\frac{3}{5}} \tag{2.62}$$

Here, C_n is the structural constant of the refractive index as a measure for atmospheric fluctuations, h is the height above ground, H_{max} the maximum height of the turbulences, and z the angular distance of the observed object from the zenith. According to Eq. (2.62), the Fried parameter is proportional to $\lambda^{6/5}$. A typical value is $r_0 = 5$–10 cm for a wavelength of 500 nm. The apertures of larger telescopes must therefore be subdivided into apertures of size r_0. Across these smaller apertures, the wavefronts are not distorted, but only tilted against the optical axis. A correction of the distortion can therefore be achieved by introducing a tilt in the opposite direction. This is usually done by subdividing one of the imaging mirrors into several smaller mirrors that are placed on tilting drives. The time constant of the

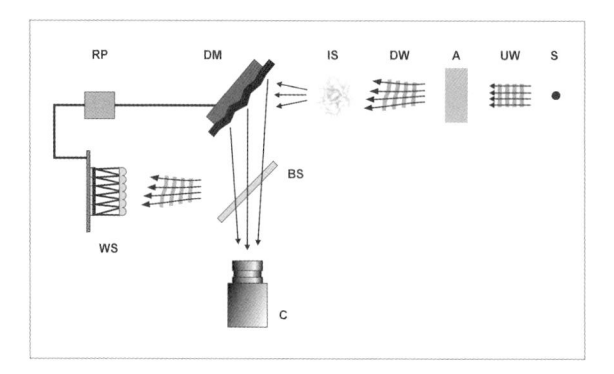

Figure 2.51 Adaptive telescope optics (schematic). S: star, UW: undistorted wavefront, A: atmosphere, DW: distorted wavefront, IS: image of the star after having passed the atmosphere, DM: deformable mirror, BS: beamsplitter, C: camera, WS: wavefront sensor, RP: real-time processor.

thermal fluctuations is of the order of some milliseconds so that the compensation circuit must be fast enough to react in realtime.

Figure 2.51 shows a schematic for a complete adaptive-optical telescope system. Light from a distant object propagates through the atmosphere and is distorted along its path. The primary mirror of the telescope picks the light up and sends it to a deformable mirror. From here, it passes a beamsplitter on its way to the telescope's camera. The light reflected by the beamsplitter is then viewed by a wavefront sensor. The beamsplitter is often coated in order to send different portions of the optical spectrum to wavefront sensor and camera. The electronic output of the wavefront sensor is finally sent to a real-time microprocessor system that calculates the signals that eventually compensate the wavefront distortions via the single mirror elements.

Wavefront sensors are also employed in the characterization of optical components or laser beam profiles. In the latter application, their enormous advantage lies in the measurement of all relevant parameters in one shot. Also, it is employed in ophthalmology for the measurement of aberrations in the human eye, as a preparation for eye surgery. For a detailed survey on the development history of the Shack–Hartmann sensor, see [97].

6.9
Determination of the Sun Angle

In recent motorcar generations, some functions and accessories meant to improve riding comfort and safety require information about the ambient brightness level. It is used, for example, for the automatic activation of the headlights, for example, in tunnels. With the advent of head-up displays (HUD) in motorcars, there are also light sensors to detect the brightness of the road directly ahead of the vehicle

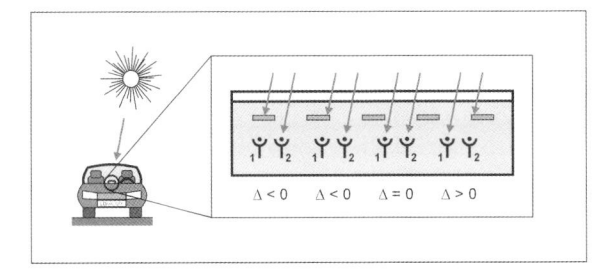

Figure 2.52 Sun angle sensor with planar design. Photodetector array with amplitude grating as shadow mask. Δ: Difference of the detector currents for each detector pair.

Table 2.35 Specifications of a sun angle sensor [98].

Property	Value
Measurement range	10°–170°
Resolution	3°
Intensity range	$1–10^6$ lx
Max. spectral sensitivity photodiodes	550 nm/650 nm
Operating temperature	−40 °C to +105 °C

to regulate the HUD's brightness. In principle, the input for all these functions is easy to acquire by a single photodetector.

Besides the integral ambient brightness level, the angle of the sun relative to the car is an important input parameter for an optimized regulation of the air condition system. The sensor is usually positioned on top of the dashboard, directly behind the windshield. It contains, depending on the sensor model, two or three photodiodes rotated in three-dimensional space to monitor different directions in space. The input signals are then combined to regulate the air flow and its direction inside the car (see, e.g., [98]).

Besides this solution, there is also an integrated sun angle sensor with a simple planar design ([99], Figure 2.52): it consists of a two-dimensional photodiode array with 64×64 detector pairs. Between the photodiode array and the entrance window of the sensor, there is an amplitude grating as a shadow mask whose grating constant is slightly different from the pitch of the detector pairs. At a given angle of the sun, the photodiodes of only one detector pair will thus receive equal intensities. On the other hand, the differences Δ of the photodiode currents will be of opposite signs on either side of this particular detector pair. Each of the 64 sensor pairs will show a zero difference for one particular angle of the sun. The angle measuring range reaches from 10° to 170°, and the angular resolution is 3° for this number of photodiodes. With its small size (Table 2.35), the sensor can also be mounted in the arm of the rear-view mirror.

6.10
Determination of Age

Ever since the advent of archeology as a scientific discipline, technical and scientific measurement and investigation methods have come in use in several fields concerning the search for artifacts and their subsequent investigation. Today, all these methods have been summarized in the science of archeometry. It is subdivided into three basic disciplines:

- Archeometric prospection covers all techniques suitable for the detection of remains under ground and helps to define the positions of possible excavation sites. Aerial photography, geomagnetic prospection, and ground-penetrating radar (GPR) are three examples for scientific prospection techniques.
- Material analysis of artifacts may give valuable information about, for example, trading routes or manufacturing technologies in ancient times. All analytical techniques that are suitable for the kind of artifact under investigation belong to this discipline.
- Archeometric dating, the determination of age of an artifact, is a very important task in the investigation of artifacts and their positions on the timescale of ancient cultures.

Particularly in the third discipline, one of the best-known techniques is radiocarbon (^{14}C) dating: it is based on the radioactive decay of natural radiocarbon that every organism takes up during its lifetime. This takeup stops when the organism ceases to exist, for example, when an animal or a person dies, or when a tree is felled. When its ^{14}C activity at this moment in time is known, and when its activity today is measured, its age, or better: the time that has passed since its death, can be determined by means of the radioactive decay law:

$$A(t) = A_0 e^{-\frac{\ln 2}{T_{1/2}} \cdot t} \tag{2.63}$$

$T_{1/2}$ is the half-life of the radioactive isotope. For ^{14}C, the half-life is well known from laboratory measurements as (5730 ± 40) years [100]. By convention, however, the half-life that is employed with Eq. (2.63) for the determination of the radiocarbon age is (5568 ± 30) years [100].

A second technique for the determination of ages between about 10^2 and 10^6 years is based on the increasing number of defects in electrically isolating crystals in time due to natural radioactivity from inside and outside the material, in particular α- and β-radiation. These radiation types have the potential to release electrons from the atoms in the crystal lattice. These electrons and the remaining, positively charged electron holes can now move freely through the crystal until they become trapped in lattice defects. Some of these traps decay in quite a short time, and their energies dissipate into the crystal lattice, which leads to a temperature increase of the crystal. Others, however, are metastable and therefore have very long lifetimes. In total, the number of occupied traps (also called "centers") increases with the time of radiation exposition until it becomes saturated.

When such a crystal structure, for example, a piece of pottery, is heated, it will emit thermal radiation. Above temperatures of 150 °C to 200 °C, however, light emissions may be observed that cannot be explained by Planck's law alone, an effect known as thermoluminescence (TL, [101]). In order to separate the TL components of the thermal background radiation, the dependence of the TL intensity on the sample temperature, the so-called glow curve, is recorded in the green or blue parts of the optical spectrum. The intensity of these emissions is proportional to the number of occupied charge traps. When D is the radiation dose, measured in Gray (Gy), and dD/dt the dose rate, given in Gray/year (Gy/a), the determination of age, Δt, then simply corresponds to the equation:

$$\Delta t = \frac{D}{dD/dt} \qquad (2.64)$$

Also, the longer the sample is exposed to radiation, the more intense the TL signal, $I(t)$, becomes until it reaches a state of saturation according to

$$I(t) \propto \tau \cdot \left(1 - e^{-\frac{t}{\tau}}\right) \qquad (2.65)$$

Here, τ is the time until about 63% of the charge traps are filled. As TL basically destroys the luminescence centers, these emissions can only be detected when the sample is heated up for the first time. At temperatures above about 500 °C, the TL of the sample will completely be quenched.

In archeometry, TL is a well-suited technique for the age determination of, for example, pottery. When a piece of pottery is manufactured, the high temperatures in the furnace completely quench the TL in the material: the dating clock is set to zero. Radioactivity will then lead to a new build-up of luminescence centers with age. As the measurements can only be carried out once and as they are destructive, only a small sample of the artifact will be used for TL dating.

Although TL seems to be a very straightforward measurement technique, its evaluation requires additional information about the radioactivity at the excavation site and of the sample itself. Excavation and sampling have to be carried out with sufficient care to avoid systematic errors. In particular, TL bleaches out when the sample is exposed to sunlight. As a result, the measurement uncertainties of ages determined with TL are around 10%, which seems to be inferior to the performance of radiocarbon dating. However, as there is no reliable calibration of ^{14}C ages older than 12 000 years before now, TL dating can compete with radiocarbon dating in this age range. On the other end of the dating range, particularly for the past 1000 years, TL is even superior due to the ambiguities of the ^{14}C ages [101]. On the other hand, due to the complexity of the sampling and measurement process, TL dating has not yet become a technique for routine analyses.

Besides TL, luminescence based on radiation defects can also be stimulated by irradiation with light. The stimulation wavelengths for this optically stimulated luminescence (OSL) depend on the material that is investigated: for feldspar,

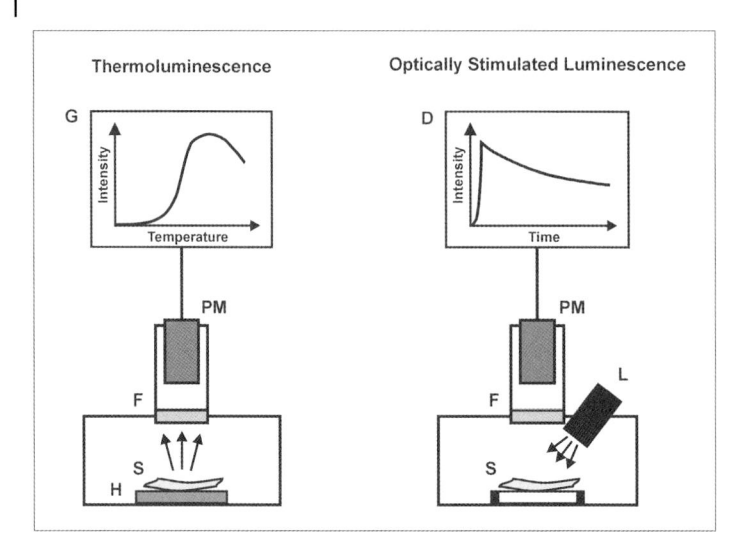

Figure 2.53 Luminescence dating techniques (schematic). H: heating, S: sample, F: filter, L: light source, PM: photomultiplier, G: glow curve, D: decay curve.

near-infrared light at around 880 nm is used, and for quartz, green light, for example, from an argon-ion laser at 514.5 nm, or filtered light from a tungsten halogen lamp. The luminescence light is analyzed at significantly shorter wavelengths, for example, in the violet/blue wavelength range for green-light stimulation, and in most of the visible spectrum for near-infrared stimulation. The measurement result is the OSL decay curve that shows a sharp increase in intensity up to a certain maximum, followed by a continuous decrease during the irradiation time. At irradiation intensities of around 10 mW/cm^2, the luminescence of feldspars and quartz will then be completely bleached out after few minutes [101]. As with TL, the sampling procedure requires safety precautions because the sample must not be exposed to light before it is analyzed.

The measurement setups for both techniques are very similar (Figure 2.53): the sample is placed in a chamber where the luminescence is stimulated either by heating it up (TL) or by irradiating it with a light source (OSL). A sensitive photodetector, usually a photomultiplier tube, picks up the luminescence signals. In the case of TL, the glow curve is recorded. In OSL measurements, the result is the decay curve.

As luminescence phenomena are based on long-term influences of natural radioactivity, their detection is a very reliable method to test the authenticity of artifacts or works of art of a certain age. As it is virtually impossible to counterfeit the TL or OSL fingerprint of an original, forgeries can generally be detected with ease. However, there are also forgeries that manage to circumvent these methods [102].

References

1. Scheibe, M.A. (2000) Quantitative Aspekte der Anziehungskraft von Straßenbeleuchtungen auf die Emergenz von nahegelegenen Gewässern (Ephemeroptera, Plecoptera, Trichoptera, Diptera: Simuliidae, Chironomidae, Empididae) unter Berücksichtigung der spektralen Emission verschiedener Lichtquellen. Dissertation. Johannes-Gutenberg-Universität Mainz.
2. Franze, K., Grosche, J., Skatchkov, S.N., Schinkinger, St., Foja, Chr. Schild, D., Uckermann, O., Travis, K., Reichenbach, A., and Guck, J. (2007) *Proc. Natl. Acad. Sci. U SA*, **104** (20), 8287–8292.
3. Hecht, E. (1989) *Optik*, Addison-Wesley, Bonn.
4. Deutsches Institut für Normung (1982) DIN 5031, Part 3. Strahlungsphysik im optischen Bereich und Lichttechnik, Beuth, Wien, Zürich.
5. Deutsches Institut für Normung (1993) DIN 1301, Part 1, Appendix A. Einheiten, Einheitennamen, Einheitenzeichen, Beuth, Wien, Zürich.
6. Hella (2005) Technical Information: Electronics–Driver Assistance Systems, Product Information.
7. Hella (2007) Technische Informationen: Fahrerassistenz-Systeme, Product Information.
8. Hagebeuker, B. (2008) *Optik & Photonik*, **3**, pp. 42–44.
9. Ringbeck, T. and Hagebeuker, B. (2007) A 3D time-of-flight camera for object detection, in *Proc. Optical 3-D Measuring Techniques, Plenary Session I, ETH Zürich*, retrieved from www.ifm-electronic.com/obj/O1D_Paper_PMD.pdf, 15.11.2009.
10. Seta, K. and Ohishi, T. (1990) *Appl. Opt.*, **29**, 354–358.
11. Agilent Corporation (2001) Agilent 5529A: Dynamic Calibrator, Product Datasheet.
12. Mikro-Epsilon (2008) optoNCDT 1401, Product Datasheet.
13. Landt, A. (2000) *Color Foto*, **6/2000**, 38–43.
14. Hexagon Metrology (2008) Leitz PMM-C, Product Datasheet.
15. Willhelm, J. (1978) Dreigitterschrittgeber. PhD Thesis. Technical University of Hanover, Hanover.
16. Dr. Johannes Heidenhain GmbH (2007) Exposed Linear Encoders. Product Catalogue.
17. Agilent Corporation (2001) Optical Mice and How They Work, White Paper.
18. Avago Technologies (2008) ADNS-5000 Optical Mouse Sensor. Product Datasheet.
19. Zeitler, R. and Berger, C. (1997) *Das Papier*, **51** (6), 287–295.
20. Haus, J. and Lauinger, N. (2007) *Laser Technik J.*, **4** (2), 39–43.
21. Haus, J. and Schaefer, R. (2003) *Technisches Messen*, **1/2003**, 10–13.
22. Bergeler, S. (2003) Einsatz optoelektronischer Flächensensoren in der ein- und zweidimensionalen Ortsfiltertechnik. PhD Thesis. University of Rostock, Rostock.
23. CORRSYS-DATRON Sensorsysteme (2008) CORREVIT S-350, Product Datasheet.

Optical Sensors: Basics and Applications. Jörg Haus
© 2010 WILEY-VCH Verlag GmbH & Co. KGaA, Weinheim
ISBN: 978-3-527-40860-3

24 MICRO-EPSILON Optronic (2008) ASCOspeed 5500, Product Datasheet.
25 ELOVIS (2006) AWS berührungslose Geschwindigkeits- & Längenmessung, Product Datasheet.
26 Ruck, B. (1990) Laser-Doppler-anemometrie, in *Lasermethoden in der Strömungsmeßtechnik* (ed. B. Ruck), AT-Fachverlag, Stuttgart, pp. 99–150.
27 Stücker, M., Baier, V., Reuther, T., Hoffmann, K., Kellam, K., and Altmayer, P. (1996) *Microvasc. Res.*, **52**, 188–192.
28 Czarske, J., Büttner, L., and Pfister, T. (2008) *Photonik*, **5/2008**, 44–47.
29 Weis, M. (1994) *Berührungslose Geschwindigkeitsmessung an festen Oberflächen mit Korrelationsverfahren*, Shaker, Aachen.
30 Merlo, S., Norgia, M., and Donati, S. (2002) Fiber gyroscope principles, in *Handbook of Optical Fibre Sensing Technology* (ed. J.M. López-Higuera), John Wiley & Sons, Inc., New York, pp. 331–348.
31 KVH Industries (2008) DSP-3000 Series, High-Performance Single Axis Fiber Optic Gyros, Product Datasheet.
32 Schreiber, U. (2000) *Ringlasertechnologie für geowissenschaftliche Anwendungen, Mitteilungen des Bundesamtes für Kartographie und Geodäsie (Band 8)*, Bundesamt für Kartographie und Geodäsie, Frankfurt.
33 Dehnen, H. (1967) *Z. Naturforschung*, **22**, 816–821.
34 www.fs.wettzell.de/. Website of the Geodetic Observatory, Wettzell of the German Bundesamt für Kartographie und Geodäsie, retrieved 15.11.2009.
35 Ouellette, F. (2001) *Spie's OEMagazine*, **1**, 38–41.
36 Kleckers, T. (2008) *Sensor Report*, **1-2008**, 14–16.
37 Zeh, T., Meixner, A., Koch, A.W., and Neumann, C. (2002) Faseroptische Bragg-Sensoren zur Dehnungs- und Temperaturmessung, in *Proceedings XVI. Meßtechnisches Symposium des Arbeitskreises der Hochschullehrer für Messtechnik* (eds W.-J. Becker and W. Holzapfel), Shaker Verlag, Aachen, pp. 65–70.
38 Micron Optics (2008) Optical Sensing Interrogator sm125, Product Datasheet.
39 Kreuzer, M. (2007) Dehnungsmessung mit Faser-Bragg-Gitter-Sensoren, www.hbm.com (accessed 15.11.2009).
40 Samfirescu, N., Wagenbach, K., Tschudi, T., Bader, M.A., Lauinger, N., and Schanze, T. (2008) A spectral filter analyzer for Bragg-grating-based strain and temperature measurements, Poster P38, in *DGAO Proceedings 2008*, Deutsche Gesellschaft für angewandte Optik, Germany, p. 126.
41 Habel, W.R., and Hillemeier, B. (1999) Rückwirkungsarme faseroptische Miniatursensoren zur Bewertung der Anfangsverformungen hydraulisch erhärtender werkstoffe, in *Fachtagung Bauwerksdiagnose – Praktische Anwendungen Zerstörungsfreier Prüfungen, DGZfP-Berichtsband 66 CD*, Deutsche Gesellschaft für zerstörungsfreie Prüfung (DGZfP), München, pp. 293–299.
42 Micron Optics (2008) Temperature Probe os4200, Product Datasheet.
43 IMPAC Infrared (2007) Series 14 / Series 15 Fast Digital Portable Infrared Pyrometers, Product Datasheet.
44 Chamberlain, J.M., Terndrup, T.E., Alexander, D.T., Silverstone, F.A., Wolf-Klein, G., O'Donnell, R., and Grandner, J. (1995) *Ann. Emerg. Med.*, **25**, 15–20.
45 FLIR Systems (2008) ThermaCAM™ E320, Product Datasheet.
46 Kamat, V. (2002) *Indian J. Anaesth.*, **46** (4), 261–268.
47 Bitmos Medizintechnik (2006) sat 800 SpotCheck Pulsoxymeter, Product Datasheet.
48 Welz, B., Becker-Ross, H., Florek, S., Heitmann, U., and Vale, M.G.R. (2003) *J. Braz. Chem. Soc.*, **14** (2), 220–229.
49 Welz, B. and Sperling, M. (1998) *Atomic Absorption Spectrometry*, Wiley-VCH Verlag GmbH, Weinheim.
50 Griffiths, P.R. and De Haseth, J.A. (2007) *Fourier Transform Infrared Spectrometry*, John Wiley & Sons, Chichester, New York, Weinheim, Brisbane, Singapore, Toronto.

51 Varian (2008) Using the Varian 670-IR Spectrometer to Observe Rotational and Isotopic Bands in CO Through High-Resolution FT-IR Spectroscopy, Application Note SI-01378.
52 Emmerich, A., Zander, K., and Seiler, W. (1991) *Zuckerindustrie*, **116**, 245–260.
53 ICUMSA (1986) *Proc. 19th Session ICUMSA*, Subj. 5, pp. 55–69.
54 Weast, R.C. (1993) *Handbook of Chemistry and Physics*, CRC Press, Boca Raton, FL.
55 ICUMSA (1974) *Proc. 16th Session ICUMSA*, Subj. 5, pp. 56–74.
56 Schmidt + Haensch (2008) Unipol L–Polarimeter, Product Datasheet.
57 Schmidt + Haensch (2005) Small Universal Polarimeter, Product Datasheet.
58 Hermann, G. and Haus, J. (1996) Polarimeters and polarization spectrometers, in *Encyclopedia of Applied Physics*, vol. 14 (ed. G.L. Trigg), Wiley-VCH Verlag GmbH, Weinheim, pp. 341–370.
59 ICUMSA (1990) *Proc. 20th Session ICUMSA*, Subj. 11.
60 LOT-Oriel Group Europe (2009) AccuMap-SE™, Product Datasheet.
61 Nimatic (2004) Brix Refraktometer, Product Datasheet.
62 Mettler Toledo (2007) Brix Measurement–Quick-Brix, Product Datasheet.
63 Krüss (2008) Refractometers, Product Documentation.
64 Helmut Hund (2002) Environmental Monitoring: TM data, TM digital µP, Product Documentation.
65 Helmut Hund (2002) Environmental Monitoring: Respicon / Respicon TM, Product Documenation.
66 Li, S.-N., Lundgren, D.A., and Rovell-Rixx, D. (2000) *AIHAJ*, **61**, 506–516.
67 Topas (2006) Laser Aerosole Particle Size Spectrometer–LAP Series, Product Datasheet.
68 OTT MESSTECHNIK (2005) OTT Parsivel®–Enhanced Precipitation Identifier and New Generation of Present Weather Sensor, Product Datasheet.
69 Haus, J. and Müller, W. (2008) P7: fluorescence microscopy made easy: routine fluorescence analysis with an LED incident-light illuminator, in: *Proc. SENSOR 2008* (ed. AMA Service), Wunstorf, pp. 151–154.
70 Precitec (2006) CHRocodile Optical Probes and Sensors, Product Datasheet.
71 Michelt, B. and Schulze, J. (2006) *Glas-Ingenieur*, **2/2006**, 35–37.
72 Sirat, G. and Psaltis, D. (1985) *Opt. Lett.*, **10**, 4–6.
73 Fries Research & Technology (2008) FRT CSL Conoscopic Holography, Product Datasheet.
74 Petter, J. (2009) P4: multi-wavelength interferometry for high-precision distance measurement, in *Proc. SENSOR 2009*, AMA Service, Wunstorf, pp. 129–132.
75 Luphos (2008) MWLI Sensorhead, Product Datasheet.
76 Dresel, T., Häusler, G., and Venzke, H. (1992) *Appl. Opt.*, **31**, 919–925.
77 Polytec (2007) TMS-300/320 TopMap In.Line, Product Datasheet.
78 Carl Zeiss Meditec (2007) Cirrus[TM] HD-OCT, Product Datasheet.
79 Mütze, K., Foitzik, L., Krug, W., and Schreiber, G. (1961) *Brockhaus ABC der Optik*, F.A. Brockhaus, Leipzig.
80 Pohl, D.W., Denk, W., and Lanz, M. (1984) *Appl. Phys. Lett.*, **4**, 651–653.
81 WITec (2006) alpha300S Scanning Near-Field Optical Microscope, Product Datasheet.
82 AT-Automation Technology (2008) C4-1280 Camera, Product Datasheet.
83 VIALUX (2008) MobilCam3D, ScanStation3D, DynaScan4D. Product Datasheet.
84 Varner, J.R. (1974) Holographic and Moiré Surface Contouring, in *Holographic Nondestructive Testing* (ed. K.R. Erf), Academic Press, New York, pp. 105–147.
85 Owen, R.B. and Sheets, P.D. (1989) *Archaeometry*, **31**, 13–25.
86 Akrometrix (2008) Model XL Production-Level Automated Flatness Inspection System, Product Datasheet.

87 Steward, E.G. (1987) *Fourier Optics – An Introduction*, Ellis Horwood Limited, Chichester.

88 Keller, G., Janschek, K., Dyblenko, S., Reimann, S., Tchernykh, V., and Trojna, G. (2004) Verbesserung der Qualitätsüberwachung für die Papierherstellung durch schnellere Online-Messung der Formation und anderer Struktureigenschaften auf der Basis eines neuartigen optischen 2D-Spektralsensors, Final Report of Research Project AiF 59ZBR, Papiertechnische Stiftung PTS, Heidenau.

89 Polytec (2007) PSV-400 Scanning Vibrometer: 1-D and 3-D Vibration Measurement, Imaging and Analysis, Product Datasheet.

90 Grosche, G., Ziegler, V., and Ziegler, D. (1985) Bronstein, Semendjajew – Taschenbuch der Mathematik, Harri Deutsch, Thun, Frankfurt am Main.

91 Dantec Dynamics (2007) 3D-ESPI System Q-300, Full-Field Measurement for Advanced Material Testing, Product Datasheet.

92 Kalms, M. and Jüptner, W. (2005) *Proc. SPIE*, **5852**, 207–213.

93 Website of Steinbichler Optotechnik; Neubeuern (Germany): www.steinbichler.de/de/main/shearagrafie__ndt.htm; website of Dantec Dynamics, Skovlunde (Denmark): www.dantecdynamics.com/Default.aspx?ID=665; website of Optonor, Trondheim (Norway): www.optonor.no/Non-destructiveTesting.aspx; website of Laser Technology, Norristown (USA): www.laserndt.com/products/home.htm, all accessed 15.11.2009.

94 Kreis, T. (2004) *Handbook of Holographic Interferometry – Optical and Digital Methods*, Wiley-VCH Verlag GmbH, Weinheim.

95 LOT-Oriel (2004) Wellenfrontsensor – Strahlcharakterisierung mit nur einer Messung, Product Datasheet.

96 Michl, J. (2008) Phänomene der klassischen Optik – Adaptive Optik, Seminarvortrag vom 12.06.2008, Universität Regensburg.

97 Platt, B.C. and Shack, R. (2001) *J. Refract. Surg.*, **17**, S573–S577.

98 Hella (2005) Electronics – Rain/Light/Solar Sensor, Technical Information.

99 Budde, W. (2008) *Automotive*, **5-6/2008**. 16–19.

100 Taylor, R.E. (2005) Radiocarbon dating, in *Handbook of Archaeological Sciences* (eds D.R. Brothwell and A.M. Pollard), John Wiley & Sons, Chichester, New York, Weinheim, Brisbane, Singapore, Toronto, pp. 23–34.

101 Wagner, G.A. (1995) *Altersbestimmung von jungen Gesteinen und Artefakten*, Enke, Stuttgart.

102 Brent, M. (2001) *Archaeology*, **54** (1), http://www.archaeology.org/0101/abstracts/africa.html (accessed 15.11.2009).

Part Three
Optics and Sensors at Work: A Laser-Spectroscopic Experiment

In industrial applications, sensors usually have a direct application to one specific measurement problem. As we have seen so far, sensor setups may be complicated, but it must be the aim of the designer of the instrument to design it as user-friendly as possible. In research applications, this is usually not the case, and experimental setups may be highly complex with a large number of optical components and sensors. This part describes an experiment that was intended to demonstrate the feasibility of a laser-spectroscopic technique for the detection of the strontium radioisotopes ^{89}Sr and ^{90}Sr in rain water [1]. It employed laser light sources, numerous optical elements, amplitude and frequency modulators, and a sophisticated detection scheme. The experiment also involved many details and principles of atomic physics and spectroscopy, so that a fully detailed description would not be within the scope of this textbook. The interested reader, however, is strongly encouraged to learn more about these details in Ref. [1].

Optical Sensors: Basics and Applications. Jörg Haus
© 2010 WILEY-VCH Verlag GmbH & Co. KGaA, Weinheim
ISBN: 978-3-527-40860-3

7
The Measurement Problem

In the wake of the Chernobyl disaster of 1986, the activity of ^{90}Sr in rain water increased from 0.4 mBq/l [2] to values of some 10 Bq/l [3]. In case of nuclear accidents, about 5% of the daughter nuclei belong to the two radioisotopes ^{89}Sr and ^{90}Sr, and their particular hazard stems from their chemistry: belonging to the same main group as calcium, these isotopes can easily be integrated into the bone tissue, once they are incorporated. As both isotopes are mere beta emitters, their radiation damages the bone marrow and may thus lead to leukemia. In addition, the long physical half-life of ^{90}Sr of 28.4 a, along with its biological half-life of up to 45 a [2], will inflict tissue damage over a long period of time.

As mentioned above, both strontium radioisotopes are pure β-emitters, and so they cannot be detected by means of γ-spectroscopy. The standard method for ^{90}Sr therefore relies on the enrichment of its radioactive daughter ^{90}Y, also a pure β-radiator. The ^{90}Sr content is then determined in the equilibrium of the cascade ^{90}Sr → ^{90}Y → ^{90}Zr (stable). This method is, however, not suitable for a rapid determination because the time until equilibrium amounts to 10 to 14 days. Also, some effort has to be taken to chemically extract the yttrium. In the years after 1986, several research groups therefore tried to find methods for a rapid test of rain water on the strontium radioisotopes, most of them based on laser-spectroscopic methods. Reference [1] reports on a feasibility study of an excitation technique that was solely based on the so-called isotope shift of atomic lines.

Optical Sensors: Basics and Applications. Jörg Haus
© 2010 WILEY-VCH Verlag GmbH & Co. KGaA, Weinheim
ISBN: 978-3-527-40860-3

8
The Physical Principles behind the Experiment

The electrons of an atom occupy discrete energy levels; their energies are quantized. When an atom absorbs a quantum of light, an electron "jumps" to a higher energy level, and the quantization of the energy levels requires that the light wavelength be tuned to match the energy difference between these two levels exactly. Figure 3.1 shows the energy level diagram of strontium with the coupled transitions and wavelengths that were relevant for this experiment: $\lambda_1 = 689.4$ nm ($5^1S_0 \rightarrow 5^3P_1$) and $\lambda_2 = 688.0$ nm ($5^3P_1 \rightarrow 6^3S_1$).

Natural strontium (with an atomic number of 38) has a relative atomic mass of 87.62 due to its composition of the four isotopes ^{84}Sr (0.56%), ^{86}Sr (9.86%), ^{87}Sr (7.02%), and ^{88}Sr (82.56%). There are slight differences in the emission wavelengths of each of these isotopes due to a shift of the atomic energy levels, the so-called level isotope shift (LIS). Suppose we had two energy levels E_A and $E_{A'}$ of two isotopes with atomic mass numbers A and A', respectively. There are three contributions to the LIS: the normal mass shift (NMS), the specific mass shift (SMS), and the field shift (FS):

$$E_A - E_{A'} = \text{NMS} + \text{SMS} + \text{FS} \tag{3.1}$$

The NMS is the result of the movement of nucleus and electrons around a common center of mass. The simple semiclassical Bohr model of an atom does not describe this effect because it assumes that the nucleus has an infinite mass as compared to the electron – the center of mass of the nucleus–electron system is thus always located in the center of the nucleus. The SMS contains influences caused by the electron collective that also result in motions around a common center of mass. Finally, the field shift is an effect that can only be understood in the framework of quantum mechanics. It stems from the fact that s-electrons have wavefunctions with nonzero amplitudes at the position of the nucleus. Thus, these electrons "see" the shape of the nuclear charge distribution. While NMS and SMS generally increase with increasing atomic mass numbers, the behavior of the FS is more complex: at a constant number of protons, an increasing mass number means an increasing number of neutrons. For the natural strontium isotopes, the neutron numbers are $N_{84} = 46$, $N_{86} = 48$, $N_{87} = 49$, and $N_{88} = 50$. In the nuclear shell model, a neutron number of 50 is a so-called magic number: all nuclear shells are occupied, and any additional neutron will lead to the occupation of a new shell,

Optical Sensors: Basics and Applications. Jörg Haus
© 2010 WILEY-VCH Verlag GmbH & Co. KGaA, Weinheim
ISBN: 978-3-527-40860-3

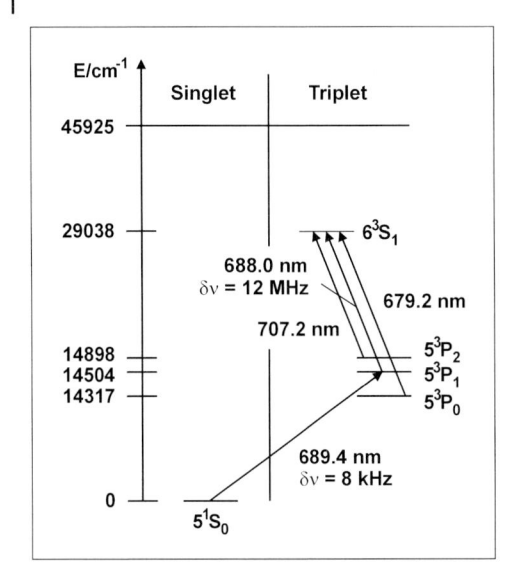

Figure 3.1 Energy level diagram of strontium with the transitions relevant for this experiment. The energy, E, is given in cm^{-1}, a common unit in spectroscopy. The inverse value of E is the wavelength of the radiation that connects the ground state with the respective excited state.

that is, to an abrupt increase in the size of the nucleus and, thus, to a sudden change in the FS. The level energy of ^{90}Sr will thus not monotonically follow the level energies of the natural isotopes.

The LIS has a direct effect onto the electronic transitions between the energy levels, leading to an isotopic shift between absorption wavelengths of the isotopes of one element. Table 3.1 lists the values for this transition isotope shift (TIS) for the two coupled transitions. This shift is usually not given in wavelength units,

Table 3.1 Transition isotope shifts of the stable Sr isotopes and of ^{90}Sr, relative to ^{88}Sr (data taken from [1] and references cited therein).

A	TIS (MHz)	
	689.4 nm	688.0 nm
84	−351.2 (18)	83.2 (41)
86	−163.7 (10)	36.6 (18)
87	−62.3 (14)	7.8 (26)
88	0.0	0.0
89	−74.8 (39)	82.5 (41)
90	−187.4 (21)	141.5 (35)

but in frequency units, and as a rule of thumb, a frequency shift of 1 MHz corresponds to a wavelength shift of 1 fm (10^{-15} m) of the visible range. The data in Table 3.1 are mostly experimental data taken from the literature cited in [1], and this reference also gives a method how the missing data were numerically calculated by interpolation. The table clearly shows that the line positions of the isotopes ^{84}Sr, ^{86}Sr, and ^{88}Sr are in monotonic arrangement, but also that the ^{90}Sr line literally "jumps" back between the line positions of ^{84}Sr and ^{86}Sr on the first transition. Also, for the isotopes with odd atomic masses, the line position is actually the center-of-mass position of a group of lines caused by the so-called hyperfine interaction between the magnetic moments of electron and nucleus (for more details on atomic physics, see, for example, [4]).

The first transition has a very low transition probability as quantum-physical selection rules actually prohibit different spin quantum numbers for initial and final atomic states. These transitions generally have small line widths and thus, the first transition is well suited to yield a spectrum in which the lines for different isotopes are well separated.

9
Spectroscopic Setups of the Experiment

The experiment was carried out in two phases. In the first, saturation-spectroscopic measurements were carried out on the first transition alone, also to verify the isotope shift data taken from the scientific literature. In the second phase, the setup was partially rebuilt to allow two-step excitations.

9.1
The Single-Step Approach

The transition isotope shift is only small compared to the linewidths that are usually measured in spectrometric setups. When performing, for example, atomic absorption spectrometry (AAS) measurements (Section 6.5.1), the isotopic shifts are generally obscured by the large linewidths involved and cannot be determined with the precision of the data given in Table 3.1. The main reason is the large linewidth of conventional light sources. This is why high-resolution spectroscopy always makes use of narrow-band laser light sources. When a laser is scanned across an absorption line, however, it will still show comparably broad line profiles. As the atoms in the sample move with a thermal velocity distribution, there will always be a velocity v_{rel} of the atoms relative to the laser beam. From the resulting Doppler effect, a relation exists between this velocity and the absorbed light frequency v (v_0: center frequency of the absorption line):

$$v - v_0 = \Delta v = \frac{v_{rel}}{\lambda} \tag{3.2}$$

Thus, the spectral width across which the laser is absorbed corresponds to the width of the velocity distribution in the (gaseous) sample. As this is described by the Maxwell distribution, the result of the laser scan will still be a Doppler-broadened absorption line. A solution to this is a technique in which the high intensity of a laser beam is employed to change the absorption properties of the sample for a second (weaker) beam. In this saturation-spectroscopic setup, the first beam, the "saturation" beam, creates such a large number of excited atoms that the absorption of the second beam, the "probe" beam, experiences reduced absorption. Pump and probe beams are usually counterpropagating, but are aligned to the same

Optical Sensors: Basics and Applications. Jörg Haus
© 2010 WILEY-VCH Verlag GmbH & Co. KGaA, Weinheim
ISBN: 978-3-527-40860-3

beam axis. Thus, the saturation effects occur with all atoms that interact with both beams simultaneously, and due to the different propagation directions, it follows from Eq. (3.2) that these are only atoms with zero velocity relative to both beams. Then, the isotope lines show up at exactly their line positions, that is, in the middle of the Doppler-broadened line profiles and with no Doppler shift. This nonlinear optical effect had not been possible to observe before the advent of the laser. For more information about this technique and other methods of laser spectroscopy, the interested reader is referred to [5]. Pump and probe beam are usually taken from one laser source, and the relative intensities are adjusted with a suitable beamsplitter. The laser source in this experiment is a dye laser running on one longitudinal mode. This type of laser uses an optical-grade jet of a dye solution, and as the absorption and emission spectra of these dyes are very broad, a dye laser is tunable over a very wide wavelength range. It delivers sufficient output power for this experiment, but it must be pumped with another high-power laser, an argon-ion (Ar^+) laser.

The sample was a strontium vapor produced by heating solid strontium with temperatures of 400 °C and above. As strontium is an alkaline-earth element, it reacts very strongly with water and oxygen. Therefore, the vapor must be kept in vacuum, so that the central part of the experiment was a vacuum furnace that contained the strontium sample.

In order to increase the signal-to-noise ratio of the detected signals, the pump beam is usually modulated. In this example, the beam is periodically interrupted by a chopper disk [6]. The probe beam intensity is then recorded with a photodiode [7], and the signal is demodulated with a lock-in amplifier whose reference is set to the modulation frequency (Figure 3.2).

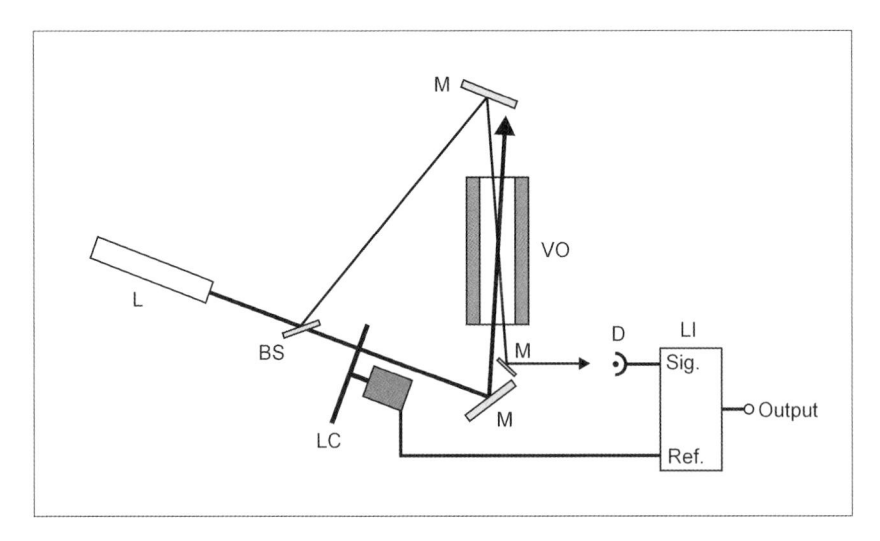

Figure 3.2 Saturation spectroscopy, schematic setup. L: laser, BS: beamsplitter, LC: light chopper, M: mirror, VO: vacuum furnace containing strontium, D: detector, LI: lock-in amplifier, Sig.: signal input, Ref.: reference input. Pump and probe beams drawn under a small angle for better visibility.

When the laser is tuned across the Sr absorption line at 689.4 nm, both lasers scan, as we have seen above, the Doppler-broadened line profile. For most of this profile, the probe beam does not experience any influence by the saturation beam, and hence, the demodulation will yield no output signal. Once, however, both beams interact with the same Sr atoms, something will happen. As we have seen above, the probe beam will experience saturated absorption by the saturation beam only when the laser is tuned exactly to the center of the absorption line. As the absorption is modulated, the lock-in amplifier will show a signal output. As this technique selects a single velocity ($v_{rel} = 0$) from the atomic ensemble, the Doppler effect will be eliminated, and the resulting line width will only show broadening influences due to collisions between the atoms and due to the laser intensities. In this experiment, however, these influences were about three orders of magnitude smaller than the Doppler width.

Figure 3.3 shows a scan across the profile of the first of the two coupled Sr transitions at 689.4 nm. As the described process works for each of the line profiles of the natural isotopes, the resulting spectrum shows all isotope lines at high resolution and with small linewidths. The figure also shows the simultaneously recorded transmission spectrum of a Fabry–Pérot interferometer (FPI, see Section 6.3) with its equidistant interference fringes. As their spacing is known from the FPI's resonator length, such a spectrum is a common high-resolution frequency reference.

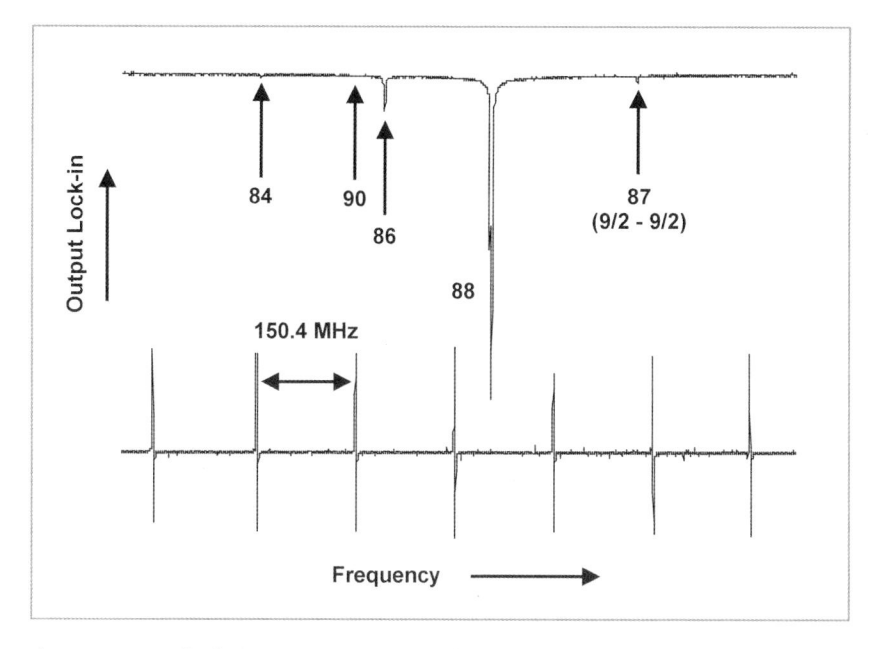

Figure 3.3 Doppler-free strontium isotope lines on the first transition of Figure 3.1, obtained with saturation spectroscopy. The numbers indicate the positions of the lines of the natural isotopes. Lower track: transmission spectrum of an FPI, recorded simultaneously. "Frequency" denotes the light frequency.

The spectrum clearly shows the lines of the natural Sr isotopes, and they are well separated from each other. Only one hyperfine component of ^{87}Sr is recorded. Their spacings correspond well to the values in Table 3.1, and the ratios of the peak heights equal those of the natural abundances of the respective isotopes. This means, for example, that the ^{88}Sr peak is about $82.56/0.56 = 147$ times higher than that of ^{84}Sr. As described above, the ^{90}Sr line will show up between the lines of ^{84}Sr and ^{86}Sr, and its height (at the post-Chernobyl rain water activities) will be so small that it becomes obscured by the two peaks of these two stable isotopes. Thus, a simple spectrometric setup involving only this very selective, but poorly sensitive transition will not be successful.

9.2
The Two-Step Approach

As a consequence of the results described in the last section, a more complicated excitation approach was devised from the energy level diagram of Figure 3.1. It involves not only the first transition, but also the transition $5^3P_1 \rightarrow 6^3S_1$ at a wavelength of 688.0 nm. This transition is not prohibited by selection rules and has a natural linewidth that is about three orders of magnitude larger than that of the first transition. This means that the excitation probability and, thus, the sensitivity are larger as well.

The basic ideas behind the new excitation schema are:

- Excite the first transition with two counterpropagating pump beams from one laser (the "pump" laser), similar to the one-step setup.
- Modulate the two pump beams with different frequencies.
- Excite the second transition with a second laser (the "probe" laser) whose beam is folded onto the axis of the two pump beams.
- Tune the probe laser to the spectral position of one of the Sr isotope lines.
- Feed the signal from the probe laser beam to a lock-in amplifier. Set its reference frequency to the sum of the modulation frequencies of the first-step transition.
- Scan the pump laser across the whole of the Doppler-broadened profile of the first transition.

The probe laser was a wavelength-stabilized diode laser built around a commercial high-power laser diode [8]. As a consequence, the detector that records the probe beam "sees" a strong absorption on the second transition only when the pump beams scan across the line of the isotope onto which the probe laser is tuned. The other isotopes are then widely suppressed. Thus, the resulting spectrum will contain the isotope lines of Sr just like in Figure 3.3, but with line heights influenced by the spectral position of the probe laser.

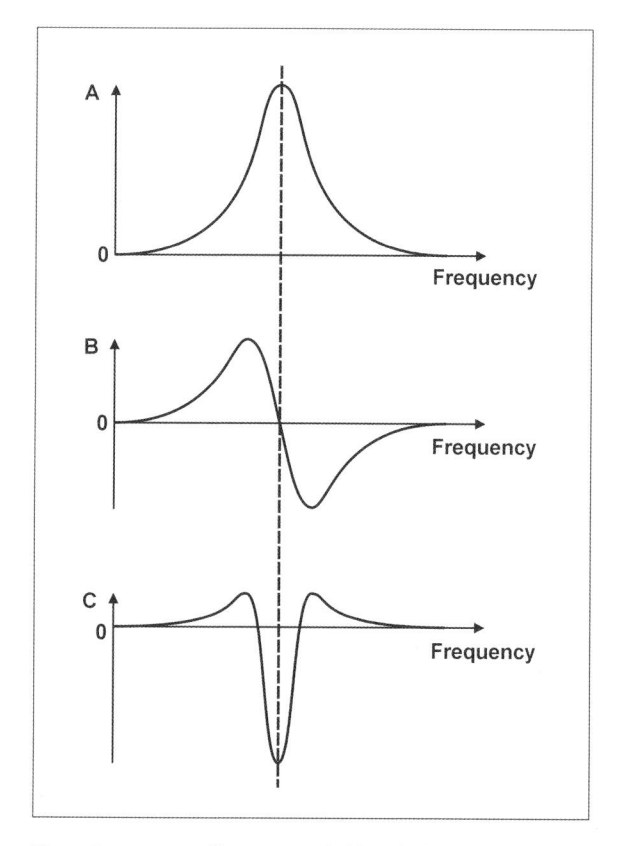

Figure 3.4 Line profiles as recorded by a lock-in amplifier. A: amplitude-modulated, B: frequency-modulated, $f_{ref} = f_{mod}$, C: frequency-modulated, $f_{ref} = 2f_{mod}$.

In order to reach even smaller line widths, and for a further reduction of the signal-to-noise ratio, one of the two probe beams is not amplitude-, but frequency-modulated. It is a basic principle of a lock-in detector that for a signal modulated with frequency f_{mod}, and for a reference frequency $f_{ref} = nf_{mod}$, $n = 1, 2, 3, \ldots$, its output during a laser scan corresponds to the nth derivative of the line profile. As an example, for $f_{ref} = 2f_{mod}$, the output will show the second derivative of the spectrum. The isotope lines in this spectrum will almost look like the lines in Figure 3.3, but with small side lobes, symmetrical to the line center (Figure 3.4).

The frequency modulation was realized with the oscillating-mirror modulator described in Section 4.7 (Figure 1.29), with a modulation frequency of 254 Hz. The mirror stroke was about 0.5 mm, with which an optical frequency stroke due to the Doppler effect of about 5 MHz could be achieved.

If f_{AM} denotes the amplitude-modulation frequency of the first pump beam and f_{FM} the frequency-modulation frequency of the second pump beam, the probe beam absorption on the second transition must then be measured on the reference frequency $f_{ref} = f_{AM} + 2f_{FM}$. This frequency is generated from the single modulation

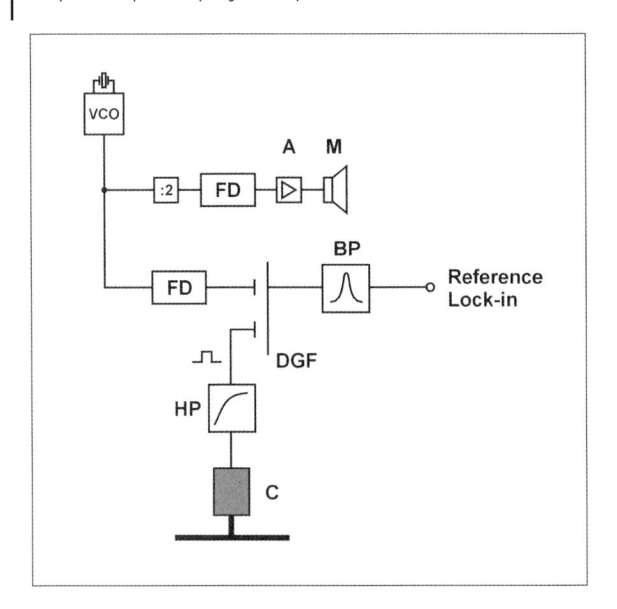

Figure 3.5 Reference frequency generation and mixing. VCO: Voltage-controlled oscillator with quartz reference, ":2": frequency divider (by a factor of 2), FD: frequency divider chain, A: amplifier, m: loudspeaker modulator, BP: bandpass filter, DGF: dual-gate FET, HP: high-pass filter, C: light chopper.

frequencies in a frequency-mixing circuit: first, f_{AM} is picked up with a light barrier from the light chopper disk. The output of the light barrier is a rectangular signal whose DC component is cut off with a high-pass filter, and which is then fed into one of the input terminals of a dual-gate field-effect transistor (FET). On the other hand, f_{FM} and $2f_{FM}$ can easily be taken from the same frequency generator and after some frequency reduction, f_{FM} drives a power amplifier and, finally the oscillating-mirror modulator. $2f_{FM}$, which is now taken as one component of the reference frequency, is fed into the second input port of the dual-gate FET. As the output of a transistor is a nonlinear function of its input, it will generate higher harmonics of the input frequencies and thus, the desired f_{ref} can be cut out of the output spectrum with a bandpass filter and then directly be employed as reference for the lock-in amplifier (Figure 3.5).

Figure 3.6 shows the schematic setup of the two-step excitation experiment. It is basically the same as for the one-step experiment, but now with the probe laser reflected onto the same axis as the pump beams. As shown in Figure 1.29, the frequency modulator also serves as a mirror that directs one of the pump laser beams into the vacuum furnace. A beamsplitter picks up the probe beam when it has passed the furnace and directs it to the detector. As pump and probe beams have linear and orthogonal polarizations, the angle of incidence onto this beamsplitter is adjusted to the Brewster angle and, thus, most of the pump beam intensity cannot enter the photodetector. The rest of it is blocked with a bandpass interference filter with a full linewidth of only 0.28 nm. Figure 3.6 shows a very simplified schematic of the experimental setup, and in its real design, quite a lot

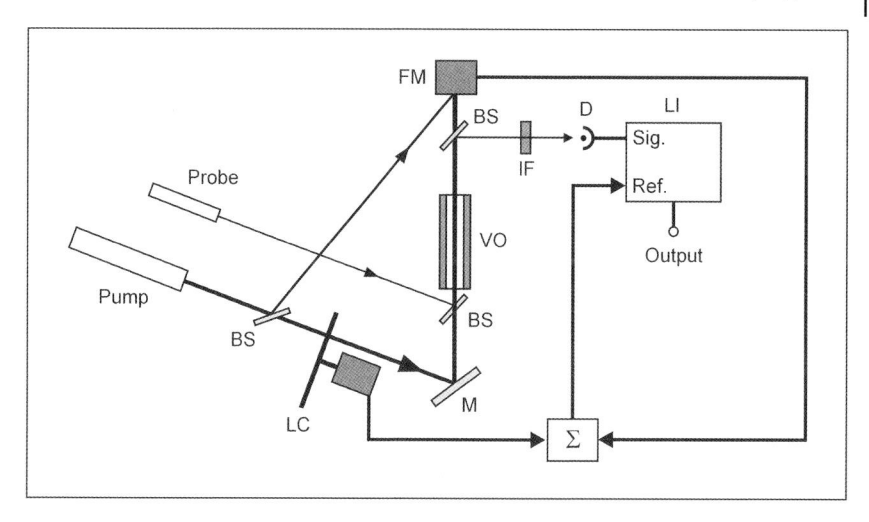

Figure 3.6 Two-step spectroscopy, schematic setup. Pump: pump laser, Probe: probe laser, BS: beamsplitter, LC: light chopper, M: mirror, VO: vacuum furnace containing strontium, FM: frequency modulator, D: detector, Σ: reference frequency generator, LI: lock-in amplifier, Sig.: signal input, Ref.: reference input.

of physical and technical implications had to be considered. Interested readers are encouraged to refer to Ref. [1] for more detailed information.

Figure 3.7 finally shows the two-step spectra obtained with the second (diode) laser consecutively tuned to ^{84}Sr, ^{86}Sr, and ^{88}Sr, and with the pump laser scanned across the profile of the first transition. It is very obvious from these spectra that the tuning of the diode laser changes the ratios between the peak heights significantly, when compared to the one-step spectrum of Figure 3.3: When the diode laser is tuned to ^{88}Sr, the line of this isotope is so dominant that none of the others shows up in the spectrum. When it is tuned to ^{86}Sr, the ^{86}Sr-to-^{88}Sr peak height ratio is about 1.4, which corresponds to an increase by a factor of about 11 relative to the ratio of the natural abundances. Tuning the diode laser to the ^{84}Sr line position, the ^{84}Sr-to-^{86}Sr peak height ratio increases by a factor of 16, and the ^{84}Sr-to-^{88}Sr peak height ratio even by a factor of 30. In addition, Figure 3.8 shows the ^{84}Sr line from the third spectrum in Figure 3.7, recorded with a high-resolution scan. It shows the typical shape of a second-derivative line with a linewidth of just a few megahertz.

Although the peak widths and the peak height ratios promise a selective detection scheme, and although the combination of a selective, but not very sensitive transition with one that is far less selective, but far more sensitive, promises an overall sensitivity that may be of an order of magnitude that is sufficient for the envisaged measurement task, the spectra still contain disturbing "stray" lines. These stray lines show up in pairs, located symmetrically around the narrow lines of the isotopes to which the probe laser is NOT tuned, and they are artifacts of two-step resonances in which only one of the pump beams is involved [1]. They can be suppressed by choosing favorable polarizations for pump and probe beams and by reducing the lifetime of the intermediate of the three involved atomic states.

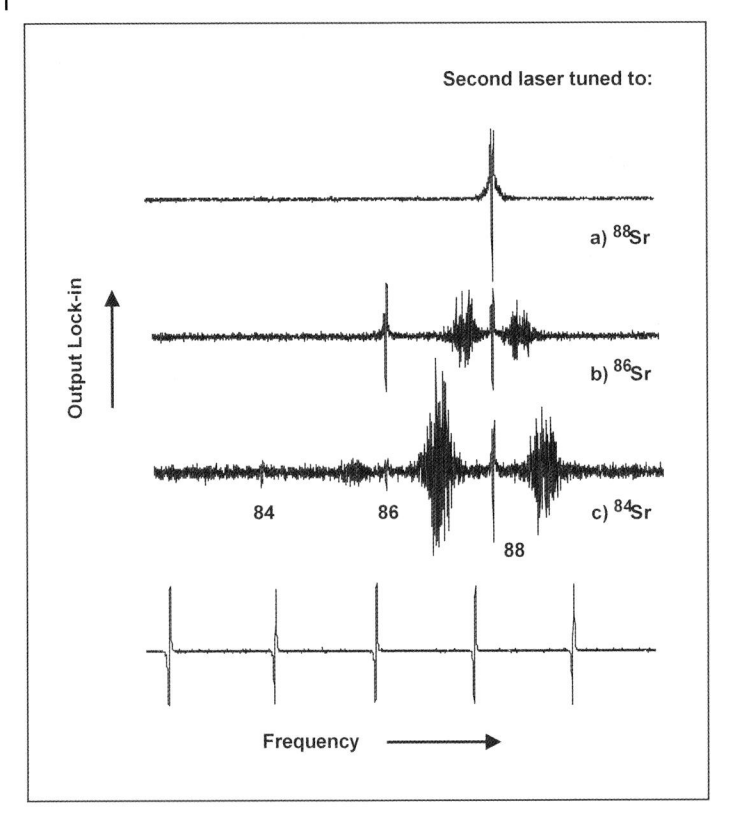

Figure 3.7 Two-step, Doppler-free strontium isotope lines with the second (probe) laser tuned to the three stable isotopes with even mass numbers. The numbers indicate the positions of the lines of the natural isotopes. Fourth track: transmission spectrum of an FPI, recorded simultaneously. The line heights are significantly changed as compared to Figure 3.3.

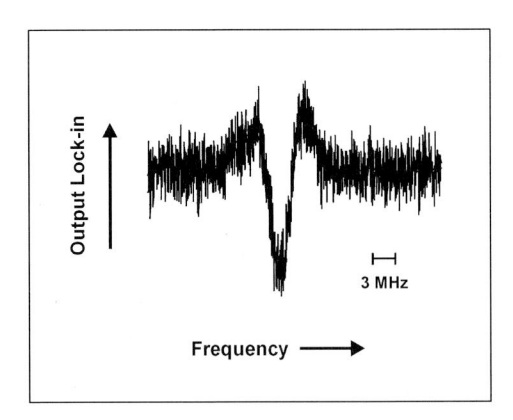

Figure 3.8 High-resolution scan of the ^{84}Sr line from the two-step spectrum of Figure 3.7, with the second (probe) laser tuned to that isotope.

In order to transform this experiment into a spectrometric detection scheme for the Sr radioisotopes, additional work would have to be done: The original idea was to chemically extract the strontium from the rain water sample and to optimize the sample preparation inside the furnace. Also, as the described experiments were carried out with natural strontium and in a vacuum furnace, the vapor density was stationary over many hours. With real-life samples, however, only very little strontium would be available for one measurement. Instead of performing a spectral scan, the lasers would then have to be kept at constant spectral positions and their absorption signal would be measured as a function of time while the furnace is heating up. This would require stabilization schemes to keep the laser lines at exactly these wavelengths, for example, by employing an additional strontium vapor cell.

References

1 Bernhardt, J., Haus, J., Hermann, G., Lasnitschka, G., Mahr, G., and Scharmann, A. (1998) Laserspektrometrischer Nachweis von strontiumnukliden, in *Zivilschutzforschung, Neue Folge Band 33*, (ed. Bundesamt für Zivilschutz), Schriftenreihe der Schutzkommission beim Bundesminister des Innern, Bonn.

2 Haberer, K. (1989) *Umweltradioaktivität und Trinkwasserversorgung*, R. Oldenbourg, München, Wien.

3 Bundesminister für Umwelt, Naturschutz und Reaktorsicherheit (1987) *Auswirkungen des Reaktorunfalls in Tschernobyl auf die Bundesrepublik Deutschland, Zusammenfassender Bericht der Strahlenschutzkommission*, G. Fischer, Stuttgart, New York.

4 Mayer-Kuckuk, T. (1985) *Atomphysik*, Teubner, Stuttgart.

5 Kneubühl, F.K. and Sigrist, M.W. (1999) *Laser*, Teubner, Stuttgart.

6 HMS elektronik (2009) HMS-220 Light Chopper, Product Datasheet.

7 Hamamatsu (2009) Si Photodiodes – S2386 series, S2386-44K, Product Datasheet.

8 Toshiba (1998) Toshiba TOLD 9150(S), Product Datasheet.

Summary

Basically, optical sensors utilize optical effects between a light source and a photodetector for the measurement of physical, chemical, and biological properties. Today, a wide variety of sources and detectors for different spectral regions exists. Light can be directed, filtered, and influenced with a large number of optical elements, and as a consequence, optical designers have a large toolbox at hand to shed first light onto their ideas for optical sensor concepts.

This, however, does not imply that the setups of optical sensors are necessarily simple. Although this is true, for example, for light barriers, an optical sensor may also incorporate quite complicated physical principles. Whatever the working principle, however, all optical sensors provide measurement data without any interference with the object under investigation. Also, they have large advantages over their nonoptical counterparts in terms of uncomplicated handling, high dynamic range, and insensitivity to electric and magnetic interferences.

The importance of optical sensorics will undoubtedly further increase, as, for example, the gathering of forces in several optical competence networks in Germany shows. With all past, recent, and future developments in mind, some say that having just passed the century of electronics with all its achievements, we have now just begun the century of the photon. It was mentioned in the introduction to this book that this collection of sensor basics and applications can by no means be complete. Although slowly becoming an everyday technique, the wide field of imaging processing techniques was not mentioned here. It is clear that they are rather a matter of image processing algorithmics than of optical technologies, but even the most sophisticated algorithms will get nowhere if either camera or the illumination of the objects to be analyzed (or both) are not adapted to the measurement problem.

Optical Sensors: Basics and Applications. Jörg Haus
© 2010 WILEY-VCH Verlag GmbH & Co. KGaA, Weinheim
ISBN: 978-3-527-40860-3

Glossary

AAS: Atomic absorption spectrometry.
Achromatic objective: Objective with correction of the chromatic aberration for two wavelengths (red/blue).
AOM: Acousto-optic modulator.
APD: Avalanche photodiode.
Apochromatic objective: Objective with correction of the chromatic aberration for three wavelengths (red/green/blue).
Bayer filter: Filter mask for two-dimensional detector arrays. Required for color detection.
Birefringence: Anisotropy of refraction.
Blackbody: Body whose spectral emissivity equals 1.
CCD: Charge-coupled device. Photodetector element for sensor arrays.
Chromatic aberration: Dispersion makes a single lens image different colors of an object into different planes.
CMM: Coordinate measuring machine.
CMOS: Complementary metal-oxide semiconductor. Photodetector element for sensor arrays.
Coherence: Describes the relative phases of light waves. If the relative phases are constant, a stationary interference pattern is visible.
Cone: Photoreceptor cell in the vertebrate retina, sensitive to color.
Correlation: Describes the similarity of two properties.
Detectivity: The inverse of the NEP.
Dichroism: Dependence of absorption on polarization.
Diffraction: Light enters areas where geometrical optics would only describe shadows.
Dispersion: Dependence of the refractive index (or diffraction angle) on the wavelength.
Dispersive element: Optical element that makes use of the effect of dispersion, usually a grating or a prism.
Doppler broadening: Broadening of spectral lines due to the Doppler effect.
Doppler effect: Frequency changes due to movements of the emission sources.
EFPI: Extrinsic fiber Fabry–Pérot interferometer.
EOM: Electro-optic modulator.

Optical Sensors: Basics and Applications. Jörg Haus
© 2010 WILEY-VCH Verlag GmbH & Co. KGaA, Weinheim
ISBN: 978-3-527-40860-3

ESPI: Electronic speckle-pattern interferometry.
ESPSI: Electronic speckle-pattern shearing interferometry.
ETA: Electrothermal atomization.
Faraday effect: Rotation of the plane of polarization by a transverse magnetic field.
FBG: Fiber-Bragg-grating sensor.
FET: Field-effect transistor.
FITC: Fluorescein isothiocyanate, a fluorochrome.
Flicker noise: Low-frequency noise with $1/f$ characteristic.
FS: Field shift, one contribution to the isotope shift.
FTIR: Fourier-transform infrared spectroscopy.
Holography: Wavefront-reconstruction technique to create 3D images.
Index of refraction: See refractive index.
INS: Inertial navigation system.
Interpolation: Technique to increase the resolution of incremental or rotary encoders.
KDP: Potassium dihydrogen phosphate, KH_2PO_4, a material used for EOMs.
Kerr effect: Square electro-optic effect, exploited in EOMs.
Lambert–Beer's law: Describes the absorption of light in a material with known absorption coefficient and thickness.
Laser: Light source based on the effect of stimulated emission.
Law of reflection: When light is reflected, the angle of reflection equals the angle of incidence.
LDA: Laser Doppler anemometry.
LED: Light-emitting diode.
Light chopper: Rotating disk with regularly spaced openings, used for modulating the amplitude of a light beam.
Light quantum: Smallest amount of light energy, calculated by multiplying the Planck constant, h, with the light frequency v.
LIS: Level isotope shift.
LOD: Limit of detection.
Longitudinal mode: Limited spectral region in which a laser emits light.
Moiré pattern: Pattern generated by the superposition of two line patterns.
Monochromator: Apparatus containing a dispersive element to select one particular wavelength.
Monomode/multimode fiber: Fiber in which only one/several modes can propagate.
MTBF: Mean time between failures.
MWLI: Multiwavelength interferometry.
NEP: Noise-equivalent power. Light power for which the signal-to-noise ratio of a photodetector equals 1.
NMS: Normal mass shift, one contribution to the isotope shift.
OLED: Organic light emitting diode.
Ommatidium: Single element of insect eyes.
Optical activity: Polarization rotation by certain substances.

OSL: Optically stimulated luminescence.

Photodiode: Semiconductor photodetector based on the internal photoelectric effect.

Photomultiplier tube (PMT): Photodetector based on the external photoelectric effect.

Plan objective: Objective with a flat image plane.

Planck's law: Describes the emission spectrum of a blackbody.

PMD: Photon-mixing device.

Pockels effect: Linear electro-optic effect, exploited in EOMs.

Polarizer: Filter that selects one state of polarization.

Polychromator: Apparatus containing a dispersive element to select several wavelengths of a spectrum simultaneously.

PSD: Position-sensitive device based on a photodiode.

Quadrature signals: Set of four periodical signals with relative phase shifts of 90°, necessary to generate directional information with encoders.

Quantum efficiency: Ratio between the number of released photoelectrons in a photodetector and the number of incident photons.

Rayleigh scattering: Scattering of particles smaller than the wavelength λ with a λ^{-4} characteristic.

Reflectance: Ratio between reflected and incident intensity.

Refractive index (of a material): Ratio between the light's phase velocity inside the material and in vacuum.

Retina: Nervous layer in the vertebrate eye that contains the photoreceptor cells: rods and cones.

Rod: Photoreceptor cell in the vertebrate retina, sensitive to low-light contrasts.

Shot noise (white noise): Frequency-independent noise.

SLR: Single-lens reflex camera.

SMS: Specific mass shift, one contribution to the isotope shift.

Snellius' law of refraction: Refraction angle and angle of incidence are linked by the refractive indices of the two materials that the light passes.

SNOM: Scanning near-field optical microscope.

Spatial coherence: Describes the area that can be coherently illuminated.

Spatial frequency: Inverse value of a structural wavelength of a surface profile.

Spatial frequency spectrum: Describes the structure of a surface.

Speckles: Granular structures in the light spots scattered by optically rough surfaces.

Spectral resolving power: Smallest wavelength spacing that can be resolved by an instrument.

Spectrum (emitted): Characteristic distribution of light intensity onto the emitted wavelengths.

Telecentric optics: Optical arrangement with a scale factor independent of the distance between optics and object.

Temporal coherence: Describes the monochromaticity of a light source. Characterized by the coherence length.

TIS: Transition isotope shift.

TL: Thermoluminescence. Employed for dating artifacts.

Transition: Electron jump between two atomic or molecular energy levels, caused by energy absorption or resulting in the emission of electromagnetic radiation.

VCSEL: Vertical-cavity surface-emitting laser.

Index

a
adaptive optics   135–138
– telescope system   138
angular velocity sensors   83–86
– earth's   85
– fiber gyroscopes   83
– laser gyroscope   85
– optical gyroscope   83
– Sagnac interferometers   83
– slip angle   83
– yaw rate   83

b
blackbody   8
– Law of Stephan and Boltzmann   10
– spectral emissivity   9
– Wien's law   9
bolometers   32

c
Chernobyl   149
chromatic confocal sensors   115–117
– chromatic aberration   116
– confocal   115
– measurement of wall thicknesses   117
CIE   7
colors   7
– perceived   7
– wavelength ranges   7
color space   7
Conoscopic Holography   117–118
contouring   123
– Moiré contouring   123
– Moiré contour map   124
– Ronchi grid   124
– Shadow Moiré   123
cross-correlation analysis   125
– transfer function   125

d
determination of age   140–142
– centers   141
– decay curve   142
– glow curve   142
– natural radioactivity   140
– optically stimulated luminescence (OSL)   141
– pottery   141
– thermoluminescence (TL)   141
2D Fourier-Transform Techniques   125–127
– 2D spatial frequency spectrum   126
– spatial frequency components   127
diffraction grating   42
– Fraunhofer diffraction   42
– spectral resolving power   42
displacement sensors   71–75
– incremental encoders   71
– Interpolation techniques   73
– optical mouse   75
distance sensors   66
– autofocus   70
– interferometric   68
– LIDAR   67
– photon-mixing device (PMD)   67
– time-of-flight   66
– triangulation   69

e
ellipsometry   102–104
– elliptically polarized light   102
– rotating analyzer ellipsometry, RAE   103
– rotating polarizer ellipsometry, RPE   103
– spectroscopic ellipsometers   104
eyes   57
– compound eyes of insects   57
–– ommatidia   57

Optical Sensors: Basics and Applications. Jörg Haus
© 2010 WILEY-VCH Verlag GmbH & Co. KGaA, Weinheim
ISBN: 978-3-527-40860-3

– cones 58
– human eye 58
– iris 58
– lens 58
– Müller cells 60
– optic disk 60
– photopic 58
– Purkinje shift 60
– retina 60
– rods 58
– scotopic vision 58

f

fluorescence detection 111–114
– filter set 112
– fluorescence *in-situ* hybridization (FISH) 114
– immunofluorescence 112
– incident-light fluorescence microscopes 114
– primary (or auto-) fluorescence 112
– secondary (or induced) fluorescence 112
– stimulated emission-depletion (STED) microscopy 114
– Stoke's rule 112

h

holographic interferometry 132–135
– double-exposure technique 134
– object beam 133
– real-time technique 134
– reconstruction 133
– reference beam 133
– time-averaged hologram 134

i

imaging detectors 33–35
– Bayer filter 35
– charge-coupled device (CCD) 33
– complementary-metal-oxide-semiconductor (CMOS) sensors 33
– detector noise 35
– – detectivity 36
– – flicker noise 35
– – noise-equivalent power 36
– – specific detectivity 36
– – "white" noise 35
– electron-multiplying CCDs (EMCCDs) 33
– fill factor 33
– Foveon X3 35
– interlaced CCDs 33
– line cameras 33
– pixels 33
– progressive-scan CCDs 33

incandescent lamps 10
– light bulb 10
– tungsten halogenide lamps 10–11
isotope shift of atomic lines 149

l

lasers 18–26
– argon-ion 26
– coherence 20
– diode 22
– dye 26
– HeNe 24
– laser threshold 18
– linewidth 20
– longitudinal modes 18
– resonator 18
– ruby 21
– safety classes 20
– semiconductor 22
– transversal modes 19
– VCSEL 24
laser vibrometers 127–129
– Bragg cell 128
– Doppler effect 127
– Laser Doppler vibrometers 127
– Mach–Zehnder-type interferometer 128
level isotope shift (LIS) 151
– field shift (FS) 151
– normal mass shift (NMS) 151
– specific mass shift (SMS) 151
light barriers 63–65
– transmission-type 63
– reflexion-type 63
Light Emitting Diodes (LEDs) 14–18
– energy gap 15
– eye safety 17
– high-power 15
– lifetime 15
– monochromatic 15
– pn-junction 15
– spectrum 16
– white-light 16
light source 5, 7
– emission spectrum 7
– emitted light power 7
– incandescent 8
– thermal 8
line sources 12
– Doppler broadening 13
– mercury 13
– mercury short-arc (HBO) lamp 14
– metal-halide lamps 14
– neon 13
– Sodium (Na) lamps 13

line spectrum 12
lock-in detector 159
– *n*th derivative 159
– reference frequency 159

m

modulators 49–52
– acousto-optical modulator (AOM) 51
– Bragg cell 52
– Doppler effect 49
– Electrooptical modulators (EOMs) 50
– Kerr cell 50
– Kerr effect 50
– light chopper 49
– magneto-optical modulators 51
– Pockels effect 50
monochromators 44
Multiwavelength Interferometry (MWLI) 118–119
– "synthetic" phase difference 118
– synthetic wavelength 119

n

Near-Field Optical Microscopy 122
– Abbe formula 122
– numerical aperture 122
– optical stethoscope 122

o

optical coherence tomography (OCT) 121
optical elements 37
– aberrations 41
– achromat 41
– apochromatic lenses 41
– distortions 41
– flip mirror 40
– image flatness 41
– lenses 40
– mirror 39
– pentaprism 40
– prisms 40
– telecentric 41
optical fibers 46–49
– acceptance angle 48
– cladding 46
– core 46
– damping 48
– disperison 48
– fiber couplers 46, 49
– gradient index fibers 48
– mode 47
– monomode fibers 47
– multimode fibers 47
– numerical aperure (NA) 48

– normal fiber frequency 47
– step-index profile 47
optical filters 43
– coatings 44
– color-glass 43
– interference filters 43
– Lambert–Beer's law 43
optical materials 37
– dispersion 37
– extinction coefficient 37
– index of refraction 37
– internal transmittance 37
– polycarbonates 38
– quartz glass 38
optical sensor 5
Organic Light Emitting Diodes (OLEDs) 16

p

particle density and particle number 106–111
– air quality monitoring 109
– disdrometers 110
– measurement of respirable dust concentrations 107
– Mie scattering 107
– particle size spectrometer 109
– Tyndall effect 107
photodetector 5
photodetectors 27–36
photodiodes 30–31
– bandwidths 30
– characteristics 30
– internal photoeffect 30
– spectral sensitivity 30
photometric units 8
– Candela 8
– Illuminance 8
– lumen 8
– luminance 8
– luminous emittance 8
– luminous energy 8
– luminous flux 8
– lux 8
photomultipliers 27–30
– channeltron 29
– dark current 28
– dynodes 27
– external photoelectric effect 27
– fall time 29
– multialkali cathodes 28
– quantum efficiency 28
– rise time 29
– single-channel electron multipliers 29
– single-photon counting 29

photoresistors 31
Planck's law 8
polarimetry 98–102
– automated circle polarimeter 100
– Biot's law 98
– chiral asymmetry 98
– circle polarimeter 100
– half-shadow polarimeters 100
– International Sugar Scale (ISS) 102
– optical activity 98
polarizers 44–46
– beamsplitter cube 45
– birefringence 45
– Brewster angle 45
– dichroism 45
– foil polarizers 45
– photography 45
– Wollaston prism 45
polychromators 44
position-sensitive devices 32
prism 42
– spectral resolving power 42

q
quantum of light 12

r
rain sensor 65
– disturbed total internal reflection 65
Rayleigh resolution criterion 42
refractometry 104–106
– Abbe refractometers 106
– degree Brix (°Brix) 105
– degree Oechsle (°Oe) 106
– Snellius' law 104
– visual handheld refractometer 105

s
saturation spectroscopy 156–158
– Doppler effect 155
– frequency reference 157
– "probe" beam 155
– "saturation" beam 155
smoke detector 64
species determination and concentration 93–114
Speckle-Pattern Interferometry 129–132
– "electronic speckle pattern interferometry" (ESPI) 130
– electronic speckle pattern shearing interferometry (ESPSI) 132
– holographic speckle pattern 129
– in-plane 131

– out-of-plane 131
– reference beam 129
– shear 132
– shearing element 132
– speckle pattern 129
– surface roughness 129
spectrometry 94
– absorption coefficient 94
– absorption spectrum 94
– atomic absorption spectrometry 96
– degree of oxygen saturation 95
– Fourier-transform infrared (FTIR) spectroscopy 97
– pulse oximetry 95
– spectral fingerprint 94
– spectrometer 94
strain sensors 86–90
– extrinsic fiber Fabry–Pérot interferometer (EFPI) sensor 89
– fiber Bragg grating (FBG) sensor 86–90
– – Bragg wavelength 86
– – k factor 88
– – optical frequency domain reflectometry (OFDR) 88
– – optical time domain reflectometry (OTDR) 88
– – time division multiplexing (TDM) 88
– – wavelength division multiplexing (WDM) 88
– resistive strain gages 86
strontium 151
– coupled transitions 152
– energy level diagram 151
– isotopes 151
– selection rules 158
– "stray" lines 161
– transition isotope shift (TIS) 152
– two-step excitation 160
– two-step spectra 161
strontium radioisotopes 149
Sun Angle Sensors 138–139
– condition system 139
– HUD's brightness 139
– regulation of the air 139
– two-dimensional photodiode array 139
superluminescent diodes (SLEDs) 17
surface topography 114–127

t
temperature sensors 90–93
– electronic pyrometers 92
– thermal FBG probe 91
– thermography 92

– thermopile detectors   92
– visual pyrometer   91
thermoelements   32
– thermopile detector   32

v

velocity sensors   75–83
– camera modules   80
– cross-correlation   77
– detector array   80
– direction of movement   79
– laser Doppler sensor   81
– time-domain cross-correlation   76
– transverse vehicle dynamics   79
vibration analysis   127–135

w

wavefront sensing   135
– artificial star (laser guide star, LGS)   136
– Fried parameter r0   136
– Kolmogorov model   136
– natural guide star (NGS)   136
– seeing   135
– Shack–Hartmann sensor   136
White-Light Interferometry   119
– broadband light source   120
– coherence radar   120
– Michelson interferometer   119